Science Skills
A level Biology

Mike Boyle

William Collins' dream of knowledge for all began with the publication of his first book in 1819. A self-educated mill worker, he not only enriched millions of lives, but also founded a flourishing publishing house. Today, staying true to this spirit, Collins books are packed with inspiration, innovation and practical expertise. They place you at the centre of a world of possibility and give you exactly what you need to explore it.

Collins. Freedom to teach

Published by Collins
An imprint of HarperCollinsPublishers
77–85 Fulham Palace Road
Hammersmith
London
W6 8JB

Browse the complete Collins catalogue at
www.collins.co.uk

© HarperCollins*Publishers* Limited 2014

10 9 8 7 6 5 4 3 2 1

ISBN-13 978 0 00 755462 1

All rights reserved. No part of this publication may be reproduced, stored in a retrieval system, or transmitted in any form or by any means, electronic, mechanical, photocopying, recording or otherwise, without the prior written permission of the Publisher or a licence permitting restricted copying in the United Kingdom issued by the Copyright Licensing Agency Ltd., 90 Tottenham Court Road, London W1T 4LP.

British Library Cataloguing in Publication Data
A Catalogue record for this publication is available from the British Library

Written by **Mike Boyle**

Commissioned by
Lucy Killick

Project managed, edited and proofread by
Cambridge Editorial

Indexed by **Chris Bell**

Production by **Emma Roberts**

Typeset by **Jouve India Private Limited**

New illustrations by **Ann Paganuzzi**

Picture research by **Amanda Redstone**

Interior design by **Anna Plucinska**

Cover design by **Angela English**

With thanks to our reviewers:
Mark Levesley, Chris Curtis, Claire Colebourn
and **Sophia Ktori**

Printed and bound by **L.E.G.O. S.p.A. Italy**

Acknowledgements

The publishers wish to thank the following for permission to reproduce photographs. Every effort has been made to trace copyright holders and to obtain their permission for the use of copyright material. The publishers will gladly receive any information enabling them to rectify any error or omission at the first opportunity.

Cover photo Wilson's Vision/Shutterstock

Fig 1a Nicku/Shutterstock, 1b Neveshkin Nikolay/Shutterstock, Fig 3 Lucky Business/Shutterstock, Shots Studio/Shutterstock, Fig 5 Nataliia Melnychuk/Shutterstock, Fig 6 luchunyu/Shutterstock, Fig 7 Martin Shields/Alamy, Fig 8 Golden Pixels LLC/Shutterstock, Fig 9 Andrey Popov/Shutterstock, Fig 10c Science Photo Library, Fig 12 Platslee/Shutterstock, Fig 21 Henrik Larsson/Shutterstock, Fig 26 deamles for sale/Shutterstock, Fig 27 Kenishirotie/Shutterstock, Fig 27 Designua/Shutterstock, Fig 29 DTKUTOO/Shutterstock, Fig 30 Michal Kowalski/Shutterstock, Fig 31 Vereshchagin Dmitry/Shutterstock, Fig 32 martynowi.cz/Shutterstock, Fig 35 a Cosmin Manci/Shutterstock, b CreativeNature.nl/Shutterstock, Fig 36 Axel Bueckert/Shutterstock, Fig 37 otsphoto/Shutterstock, Fig 38 foto-ann/Shutterstock, Fig 39 Photok.dk/Shutterstock, Fig 40 sar_38/Shutterstock, Fig 41 Ana-Maria/Shutterstock, Fig 42 Chin Kit Sen/Shutterstock, Fig 43 Vit Kovalcik/Shutterstock, Fig 45 Lightspring/Shutterstock, Fig 46 CLIPAREA l Custom

Contents

Getting the best from the book 1

SKILLS

Working Scientifically

- **S1** Science today: how science works, then and now 4
- **S2** Theories, idea and hypothesis: what is a theory? 4
- **S3** The scientific method: investigating cause and effect 5
- **S4** Testable ideas: the experimental hypothesis; the null hypothesis 6
- **S5** Control experiments: how to validate your investigation by making effective comparisons 7
- **S6** Investigations involving people: control groups; randomisation 7
- **S7** The placebo effect: what it is and how it affects investigations and trials 8
- **S8** Blind, double-blind and open label trials: eliminating the placebo effect 8
- **S9** Drug development: the vital steps in getting a drug to market 9
- **S10** Clinical trials: testing on real patients 10
- **S11** Sharing the knowledge – scientific papers, journals and peer review: What's in a research paper? 10
- **S12** Accuracy, precision, reliability and validity: related to collecting data 12
- **S13** Random and systematic errors: sources of inaccurate measurements 12
- **S14** Drawing valid conclusions: deciding if a conclusion is valid 13
- **S15** Hazards, risks and risk assessments: who's responsible?; what to look out for 13
- **S16** Models and modelling: how models help; Computer models 14

Quality of Written Communication

- **S17** Writing for your intended audience: thinking about purpose, audience and format 14
- **S18** Ensuring meaning is clear: the importance of legible text, correct spelling, accurate punctuation and grammar 15
- **S19** Organising information clearly and coherently: writing long answers and synoptic essays 15
- **S20** Using specialist vocabulary: using advanced level terminology 16
- **S21** Command words in exams: understanding what examiners expect 17

Maths

- **S22** Units of size in biology: the three units you will need 19
- **S23** Biological units and standard form: a simple way to work with large or small numbers .. 19
- **S24** Units of volume and weight: using SI units; the central role of water 20
- **S25** Microscopes and magnification: actual size, magnification and observed size 20
- **S26** Percentages and estimates: calculating without a calculator; the power of guesstimates 21
- **S27** Ratios: the two ways of writing ratios 22
- **S28** Mean, median and mode: what they are and when to use them 22
- **S29** Percentiles, deciles and quartiles: what they are and when to use them 23
- **S30** Working with concentrations: molarity and percentage 24
- **S31** Water potential: why osmosis happens; Deciding which way water will move 25
- **S32** How to construct a results table: the golden rules 26
- **S33** What to plot? Different types of graph: Deciding which graph to draw 26
- **S34** How to plot a line graph: the four steps 28
- **S35** Reading graphs: Interpolation and extrapolation 28
- **S36** Correlations and scattergrams: what makes a good correlation 28
- **S37** Logarithmic scales: plotting a wide range of numbers on one axis 29

S38	Sampling: how to get good data without wasting too much time 29
S39	Hardy-Weinberg equilibrium: looking at frequencies of alleles in populations; finding out if populations are evolving 30
S40	Statistical tests and values of p: how to determine whether results are significant or not 31
S41	Choosing the right statistical test: simple flow chart to decide which test to use 31
S42	Normal distribution and standard deviation: the spread of data about the mean 32
S43	Standard error with 95% confidence limits: using graphs to show the difference between sets of data 33
S44	The chi-squared test: comparing observed with expected when the data falls into categories 33
S45	Statistics – the t-test: using mean and standard deviation to compare sets of data 34
S46	Statistics – the Spearman Rank test: how close is the correlation between two sets of ranked data? 34

Skills to Activities table **35**

ACTIVITIES

A1	Developing a new oral rehydration therapy 37
A2	Testing a new fertiliser 38
A3	How clean is your river? 40
A4	The discovery of viruses 42
A5	How is diabetes diagnosed? 44
A6	How quickly can your body cope with sugar? 45
A7	Parasites 46
A8	Pedigree cats 48
A9	Who's hiding a recessive allele? 49
A10	Pesticides, pollutants and food chains 50
A11	Can we control an animal's growth rate? ... 52
A12	Who's at risk of heart disease? 55
A13	The potential of stem cells 57
A14	Where have all the UK's forests gone? 59
A15	Overweight or just big-boned? 61

ASSESSING INVESTIGATIVE SKILLS

AIS1	 64
AIS2	 66
AIS3	 69

ANSWERS

A1	Developing a new oral rehydration therapy 71
A2	Testing a new fertiliser 71
A3	How clean is your river? 72
A4	The discovery of viruses 72
A5	How is diabetes diagnosed? 73
A6	How quickly can your body cope with sugar? 74
A7	Parasites 74
A8	Pedigree cats 74
A9	Who's hiding a recessive allele? 75
A10	Pesticides, pollutants and food chains 75
A11	Can we control an animal's growth rate? ... 76
A12	Who's at risk of heart disease? 76
A13	The potential of stem cells 77
A14	Where have all the UK's forests gone? 77
A15	Overweight or just big-boned? 78
AIS1	 78
AIS2	 79
AIS3	 80
QWC Worked Examples 82

Glossary 85
Appendix International System of Units (SI) 89
Index 90

Getting the best from the book

We have designed this book to give you all the support you need to master the key skills necessary for success on your course. The science, maths and quality of written communication skills, for every major exam specification, are explained in detail. There are activities to practise all the skills so you have a chance to apply your learning and they are set in interesting contexts to give you the chance to use your skills in unfamiliar situations. We have included the answers so you can check your understanding and also see how you can improve using the helpful hints, tips and pointers. The Assessing Investigative Skills activities give you the opportunity to practise some of the skills you will require in assessed practical tasks. To help you improve the quality of your written communication, we have included worked examples to show how low, medium and high level answers get their marks. The glossary will help you to check your understanding of key terms.

SKILLS

Each skill is colour coded to tell you what type of skill it is:
Working Scientifically
Quality of Written Communication
Maths.

Each skill is explained in detail.

The activities that practise the skill are listed at the end of the skill.

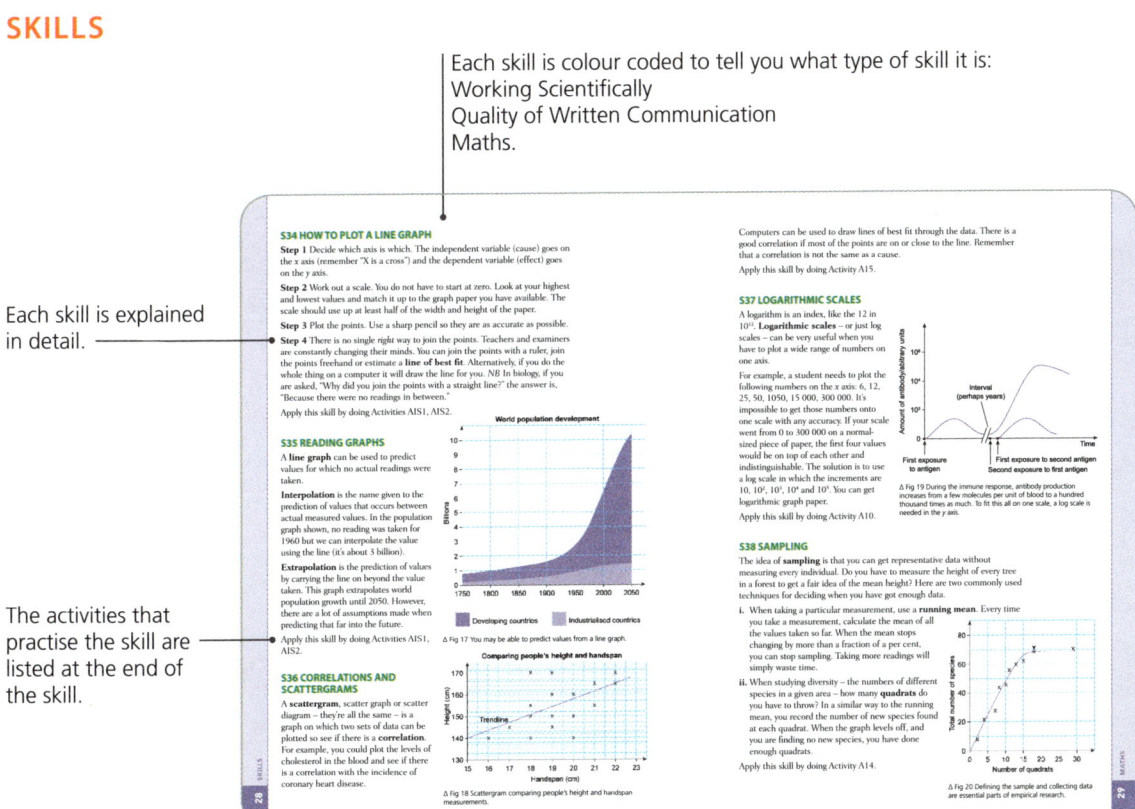

ACTIVITIES

The activities are listed in a logical sequence to allow you to tackle increasingly complex ideas as your knowledge of biology deepens.

A7 PARASITES

Statistics: using the Spearman rank test

In parasitic relationships, one species – the parasite – benefits in some way while the other – the host – suffers some harm to a greater or lesser extent. It is a harsh but simple biological fact that the vast majority of wild animals are infested with parasites. Fleas, tics, intestinal worms – it seems that all large organisms are under attack from smaller ones that want a piece of them. Parasites evolve with their hosts so that in most cases a host can carry a certain population of parasites without showing any major ill effects. So is it true that the larger the organism, the more parasites it can carry?

The Spearman rank test is used to assess the degree of correlation between two ranked sets of data. Ranked data has an *order* to it: biggest to smallest, lightest to heaviest, and so on. So Spearman rank is suitable for use on data resulting from hypotheses like *Students who are good at science will also be good at maths* or *The more fat in the diet, the higher the incidence of heart disease.*

Δ Fig 35 Almost all wild animals have parasites, both inside and outside their bodies. The flea is a common parasite and most individuals can support a population of fleas without showing any ill effects.

QUESTIONS

In this example, a researcher investigated the hypothesis *The larger the mouse, the more fleas it will have.*

Step 1 Collect your data

You would gather two sets of data. Data set 1 is the weight of the mouse and data set 2 is the number of fleas on each individual mouse.

1 Weight of mouse (g)	2 Number of fleas	3 Rank of data set 1	4 Rank of data set 2	5 *d* (column 3 minus column 4)	6 d^2
15.3	12				
18.0	18				
21.5	22	1	1		
20.2	19				
16.7	14				
17.9	15				
19.8	17				
19.0	18				
18.8	20				
17.2	13				

Δ Table 13

Step 2 Rank the data

The heaviest mouse is rank number 1 and the greatest number of fleas is rank number 1. As there are 10 measurements, the last rank is 10.

1. Fill in the missing values in columns 3 and 4.

46 A7 PARASITES

Step 3 Work out values of *d* and d^2

2. Work out the values in column *d*, where *d* is the difference between the rank in data set 1 and the rank in data set 2.
3. Work out the values of d^2.
4. Suggest the advantage of using d^2 over *d* when working out the Spearman rank value.

Step 4 Substitute this data into the formula for Spearman rank

$$r_s = 1 - \frac{6\sum d^2}{n(n^2 - 1)}$$

Where

r_s is the Spearman rank, given the name rho (pronounced row, as in "row the boat")

n is the number of data sets you have.

5. Work out the value of r_s.

Interpreting the value of r_s

All values for Spearman rank fall between −1 and +1. A value of +1 means a perfect positive correlation – in this case, the larger mouse always has more fleas – and a value of −1 means a perfect negative correlation. A score of −1 here would mean the smaller the mouse, the more fleas it carries.

As a rough guide, the value can be placed on a scale like this, and the more towards an extreme your values lies, the better the correlation.

−1 ________________ 0 ________________ +1

However, looking at a value's position on a line is only a rough guide. To be more specific, we can get a value of p by looking our value up on a table of critical values (Table 14).

n	0.10	0.05	0.025	0.01
4	1.000	1.000	–	–
5	0.800	0.900	1.000	1.000
6	0.657	0.829	0.886	0.943
7	0.571	0.714	0.786	0.893
8	0.524	0.643	0.738	0.833
9	0.483	0.600	0.700	0.783
10	0.455	0.564	0.648	0.745
11	0.427	0.536	0.618	0.709
12	0.406	0.503	0.587	0.678

Δ Table 14 Critical values for Spearman rank values

QUESTIONS

6. What is *n* in our example? (It might help to highlight the relevant row.)
7. Our Spearman rank value exceeds the critical value in the 0.01 column. What does this mean?
8. QWC Outline the differences between a mutualistic and a parasitic relationship. Illustrate your answer with examples.

Skills practised

18–22, 47

A7 PARASITES 47

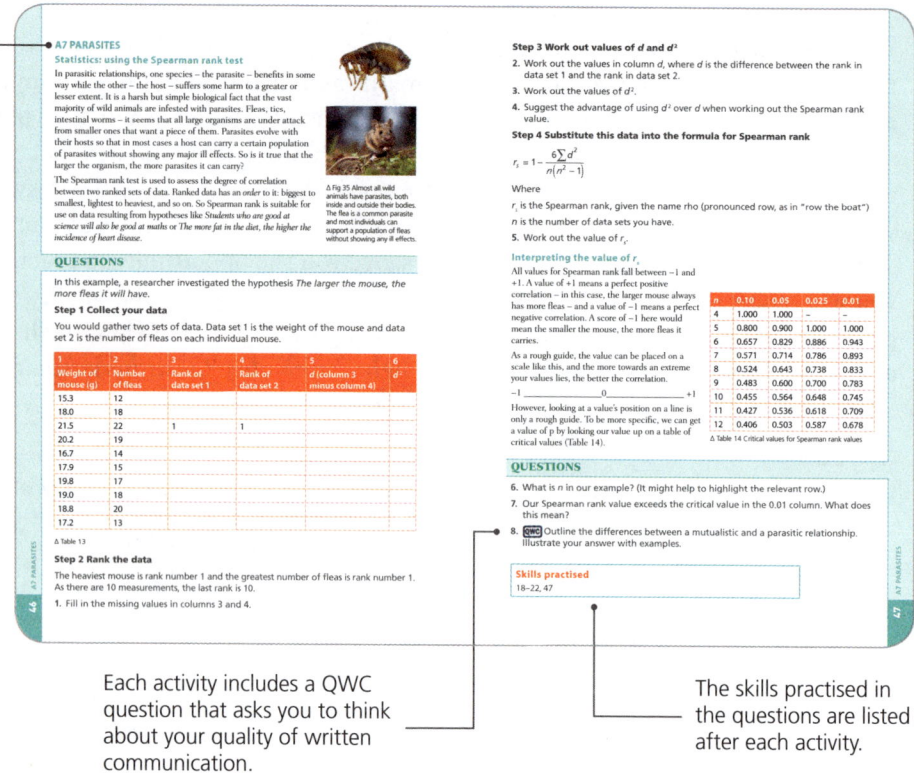

Each activity includes a QWC question that asks you to think about your quality of written communication.

The skills practised in the questions are listed after each activity.

There are three Assessing Investigative Skills activities.

Assessing Investigative Skills

AIS1

Investigating the effect of temperature on the activity of trypsin

Trypsin is a protein-digesting enzyme made by the pancreas. It forms part of pancreatic juice, which is secreted on to food in the duodenum. Trypsin's function is to hydrolyse protein molecules into shorter chains of polypeptides.

To investigate how trypsin activity can be affected by temperature – or any other variable – we need to give it some protein to digest. Egg white is almost pure protein. Small glass tubes, open at both ends, can be filled with raw egg white and then boiled so that the protein solidifies. When dipped into enzyme solution the enzymes will digest the egg white from each end. The faster the enzyme activity, the faster the egg white will disappear.

A student chose a range of temperatures to investigate and wrote down the results (Table 22).

Temp (°C)	Time taken to digest protein (min)	Rate (1/time taken) × 100	Rate × 100
10	83	0.012	1.2
20	41	0.024	2.4
30	22	0.045	4.5
40	12	0.083	8.3
50	35	0.028	2.8
60	(protein not digested)	-	-

Δ Table 22

In this investigation:

1. What was the independent variable? (1)
2. What was the dependent variable? (1)
3. List three variables that would need to be controlled. (3)
4. Suggest a suitable control. Explain your choice. (2)
5. Explain why this was a suitable range of temperatures. (2)
6. Explain why a line graph was suitable to display these results. (1)
7. A simple way to work out the rate or reaction is to do the reciprocal of time, that is, 1/time. Suggest why *1/time* is better than *time* as a measurement of enzyme activity. (1)
8. Suggest a simple practical reason why rate was multiplied by 100 in the final column. (1)

64 ASSESSING INVESTIGATIVE SKILLS

Δ Fig 55

9. We can use the graph to estimate the rate of reaction at temperatures for which we took no readings.
   a) What is this process called? (1)
   b) Estimate the rate of reaction at 25 °C. (1)
10. We can also estimate readings by following the line of the graph beyond the range of the readings.
    a) What is this process called? (1)
    b) Predict the rate of reaction at 55 °C. (1)
11. The student concluded that the optimum rate of reaction is at 40 °C. Is this a valid conclusion? Explain. (1)

The temperature coefficient, known as Q_{10}, is a measurement of the effect of a 10 °C rise on the rate of enzyme activity.

The formula for Q is: $\frac{\text{rate of reaction at t + 10 °C}}{\text{Rate of reaction at t °C}}$

12. Work out the Q_{10} when t is 20 °C. (2)
13. Relate the properties of enzymes to the fact that they are proteins. (7)

Skills practised

3, 5, 18–22, 34, 35, 36

ASSESSING INVESTIGATIVE SKILLS 65

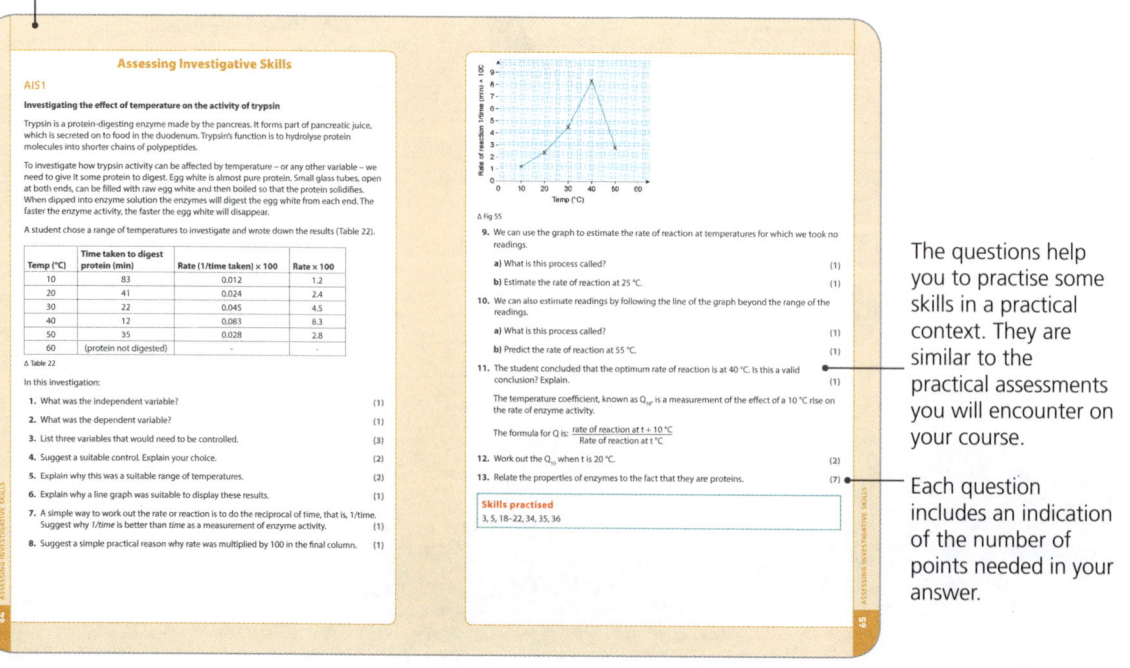

The questions help you to practise some skills in a practical context. They are similar to the practical assessments you will encounter on your course.

Each question includes an indication of the number of points needed in your answer.

ANSWERS

The answers include helpful tips and hints.

Each QWC answer includes low, medium and high level answer pointers to help you improve.

Fully worked-out approaches are included to help you build up the skills you require.

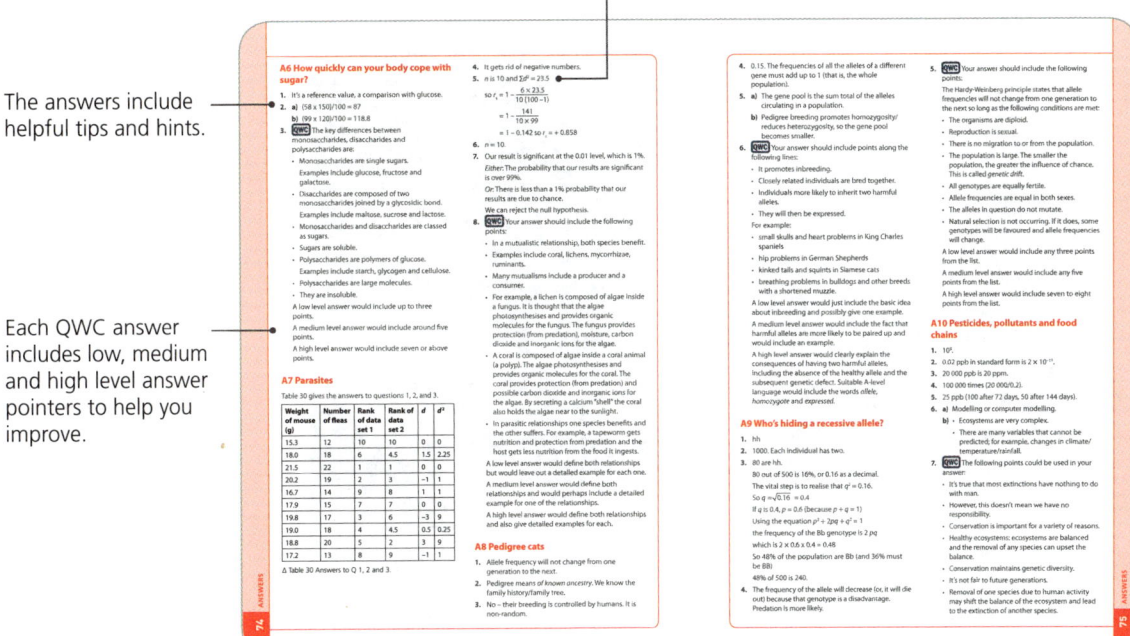

The QWC Worked Examples show low, medium and high written responses.

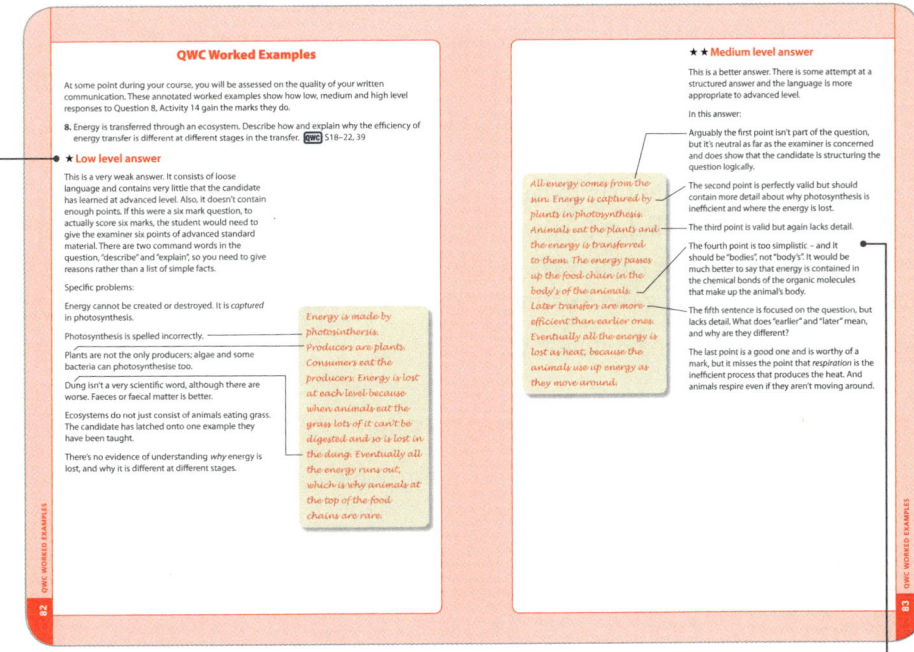

Comments explain how to improve the quality of written communication.

Skills

S1 SCIENCE TODAY

△ Fig 1(a) Charles Darwin (1809–82).

△ Fig 1(b) Marie Curie (1867–1934).

Ask anyone to name a famous scientist and they will almost certainly name someone who is dead. The likes of Newton, Faraday, Curie, Darwin, Einstein and many others achieved fame because they made major breakthroughs. They were geniuses who, to quote Albert Szent-Gyorgyi, "saw what everyone has seen before, but thought what nobody had thought". It's interesting to speculate that there must have been many anonymous scientists who wasted their time, pursued mad theories, poisoned themselves or blew themselves up.

It's tempting to think that progress in science today is similar, but it isn't. Today there may well be more scientists alive and working than all of the dead ones put together. Scientists today must be highly skilled at building on the existing bank of knowledge that we already have and developing that according to strict rules and ways of thinking. During your course, you must practise acknowledging the existing knowledge in your work.

Apply this skill by doing Activity A4.

S2 THEORIES, IDEAS AND HYPOTHESES

"No amount of experimentation can ever prove me right; a single experiment can prove me wrong" – Albert Einstein.

A famous newsreader, hosting a TV debate about evolution versus creationism, started by saying to the scientist: "Evolution – it's only a theory, isn't it?"

There is a widespread belief that theories are debatable and facts are, well, facts. However, to a scientist, a **theory** is a conceptual framework that is used to explain what we already know, and to make predictions.

Science starts with observations, followed by questions. People observe the world and say "could it possibly be that…". If this idea can be tested, it's called a hypothesis. If it can't, it isn't.

Generally, science proceeds by trying to prove hypotheses wrong. This is a much more effective approach than trying to prove hypotheses correct. A vital point here is that if we cannot disprove a hypothesis, we do not prove it right, but we *gather support* for it. All of the scientific knowledge we have today consists of the best hypotheses that have gathered a lot of support and have not been disproved.

So it's true that evolution is a theory, but so is the idea that everything is made from atoms. Both are very good theories because there is lots of evidence to support the idea, and none – that are scientifically credible – to disprove them.

Apply this skill by doing Activity A4.

S3 THE SCIENTIFIC METHOD

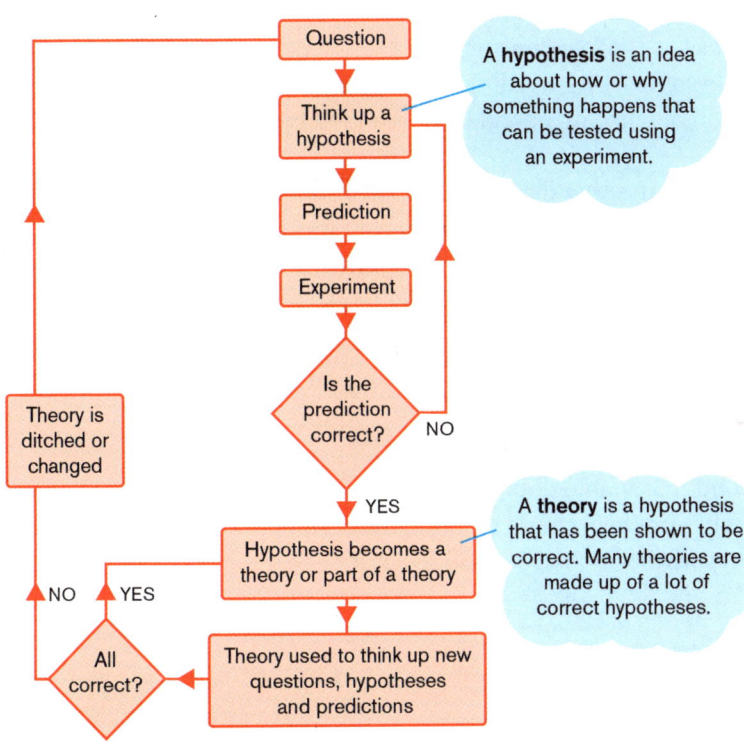

△ Fig 2 The scientific method.

Basically, science is about investigating cause and effect: what causes what. The scientific method is a logical way of going about this process, by eliminating the possibility that what you are looking at – the effect – could be caused by anything else.

The scientific method is basically a "thinking toolkit" that allows us to draw valid conclusions.

- The process starts with a hypothesis, which is a testable idea.
- Then you vary one factor – the **independent variable**. That's the cause.
- And you measure its effect on another factor – the **dependent variable**. That's the effect.
- All other variables should be controlled, that is, kept the same. If we do not do this, we cannot say that our specific cause is responsible for the effect.

Apply this skill by doing Activities A1, A2, AIS1, AIS2, AIS3.

Biological investigations can be difficult, however, because, living things are complex and it isn't always possible to control all the variables.

Apply this skill by doing Activity A1.

S4 TESTABLE IDEAS

A vital starting point for the scientific method is the hypothesis – an idea you can test. There are two basic types:

- The **null hypothesis** is the default position of the investigation. The wording of the null hypothesis could be something like, "Light intensity has no effect on the distribution of algae." The ability to reject the null hypothesis is absolutely central to the scientific method.
- A vital partner to the null hypothesis is the *experimental hypothesis*, sometimes called the *alternative hypothesis*. Its wording would be something along the lines of: "The distribution of algae in a pond is determined by light intensity."

So, to see which hypothesis to support, you gather data and perform a suitable statistical test – the exact one depends on the nature of the investigation – and then you can decide whether to *accept* or *reject* the null hypothesis. By rejecting the null hypothesis, you *gather support* for your experimental hypothesis. However, you can never say that you have *proved* your hypothesis. If you accept the null hypothesis, the experimental hypothesis is rejected.

A good approach to understanding the null hypothesis is to compare it to a trial in which a defendant is innocent until proven guilty. The default position is *not guilty* until we are sure, beyond reasonable doubt, that the defendant is *guilty*. It's the same with science. The null hypothesis is a not guilty approach. If it proves unlikely that the null hypothesis is true, we will reject it.

A word of warning here. Although it is philosophically correct to say that you can never prove anything, in the real world this approach has its problems. Sometimes the evidence is so overwhelming that if you assume that you *have* proved something, you won't go far wrong. For example, it has never been proved that smoking causes lung cancer, a fact seized upon by the tobacco companies in their defence. However, if you accept the link as "case proven, cigarettes are guilty", you'll probably have a longer and healthier life.

Apply this skill by doing Activities A11, AIS3.

△ Fig 3 Most people accept there is a link between smoking and lung cancer.

S5 CONTROL EXPERIMENTS

Control experiments are *comparisons*. They show that the results obtained must be due to the variable under test, by showing what would happen without that factor.

For example, in an investigation into the effect of heat on enzyme activity, you might vary the temperature and record the time taken for a protein-digesting enzyme to break down a piece of egg white. How do you know that it is the enzyme that is digesting the protein and nothing else, such as the heat itself? In this case the **control** investigation would be to repeat the whole investigation using boiled enzyme, or no enzyme at all. If the protein remains undigested at all temperatures, logic tells you that the results must be due to the enzyme.

When working with living things, however, controls can be difficult. It's not too bad if you are working with plants: if you have a new fertiliser to test you could quite easily start with two crops of identical plants, grow them in the same conditions, and then give the experimental crop the fertiliser while the other is the **control group**. When you've finished, you can put them all in the oven to dry them out and see which group, experimental or control, had gained the most mass.

But you cannot do this with people. Everyone is different and it is impossible to control all the variables. So let's look at how you go about working with people.

Apply this skill by doing Activities A2, A11, AIS1.

△ Fig 4 Adding fertiliser to one batch of plants.

S6 INVESTIGATIONS INVOLVING PEOPLE

Trials involving people take place more often than you might think; they are the only way to trial new drugs effectively. But how can you be strictly scientific? In an ideal world, you would take 100 pairs of identical twins and split them into two groups, but there are a limited number of identical twins and they probably wouldn't be too happy about volunteering for lots of trials.

The solution is to get as many people as possible – for some trials this may involve thousands – and split them into two groups that are as identical as possible. We call this **randomisation**, and it is important because it eliminates selection bias, and therefore makes sure that the trial is a fair comparison. The two groups need to be similar in **prognosis**, meaning that without treatment the two groups would be expected to fare equally well. If the two groups are properly randomised, any differences are likely to result from differences in treatment.

Randomisation can be done very effectively on a computer. However, the smaller the group the greater the probability that chance produces unbalanced groups. To overcome this, there can be *block randomisation* to ensure that, say, each group gets equal numbers of males and females.

Two properly randomised groups would need to contain:
- the same range and balance of ages
- the same balance of males and females (unless trialling a sex-specific treatment, such as a contraceptive).

They also need to be matched for their general health. If you are trialling a preventative medicine such as a vaccine, you need all the people to be free from the disease at the start. If you are trialling a cure, such as a statin that lowers blood cholesterol levels, you need all the volunteers to have high cholesterol so that you can clearly see the effect of the new drug.

Apply this skill by doing Activities A1, A13.

S7 THE PLACEBO EFFECT

The **placebo effect** is a measurable, observable or felt improvement in a patient's health that is not due to a medicine or invasive procedure (such as surgery). As a simple example, a patient can go to the doctor complaining of headaches, be given a completely useless pill containing sugar or chalk (that is, a **placebo**) and feel much better afterwards.

It has been said that the placebo effect is simply a case of mind over matter – a result of the patient's faith in the doctor, or in the power of medicine – but it can result from a number of other causes, such as a condition that would have got better on its own, or a misdiagnosis in the first place.

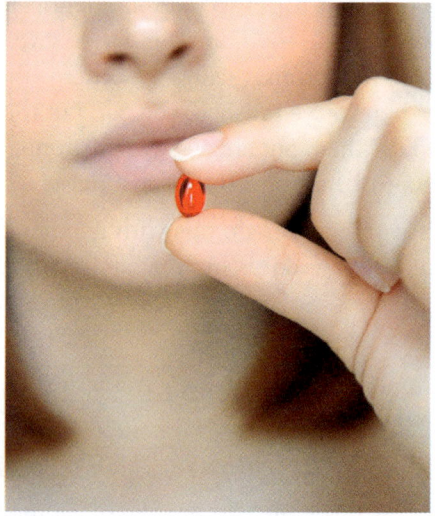

△ Fig 5 Is she getting the new drug, or simply a sugar pill? The patient shouldn't know, and neither should the experimenter who gave her the pill.

The placebo effect is a very useful tool in medicine, but it can make a complete mess of drug trials. How do you know if it's the new drug that's having the effect, or simply the thought of being given a wonderful new medicine? The solution is a blind trial. You need to be able to outline trials for new drugs and vaccines, and the placebo effect is an important consideration.

Apply this skill by doing Activity A1.

S8 BLIND, DOUBLE-BLIND AND OPEN LABEL TRIALS

The purpose of blind and double-blind trials is to eliminate the placebo effect.

In a simple blind trial, patients do not know if they are getting the active drug or not. However, it has been found that if the health workers giving the treatment *know* that they are giving the active drug, they will act differently. They give off subtle cues that the patient picks up, and this also brings about the placebo effect. For this reason, the double-blind trial was developed.

In a double-blind trial neither the health worker nor the patient knows whether they are receiving the active drug or not. Only the people who prepared the medicine know which is which. If the two groups have been properly randomised (see S6 Investigations involving people), researchers should get a clearer picture as to whether the drug works or not.

In contrast, in **open label trials** all parties know exactly which treatment they are getting. Sometimes there is no alternative to an open label trial. For example, in a trial that compares an anti-inflammatory drug with a physical therapy such as ultrasound or massage, you wouldn't need to be a genius to figure out which treatment you were being given.

Open label trials are commonly used as part of **clinical trials** (see below). Once a drug's effectiveness has been established, open label studies can give very useful information about side effects and long-term safety. Open label trials are so called because they contrast with blind studies, in which the labels on the drug boxes have been covered up.

Apply this skill by doing Activity A1.

S9 DRUG DEVELOPMENT

Getting a new drug to market is a long process. From its discovery or manufacture it can take over 10 years before a drug is awarded a licence and then can therefore be prescribed or sold over the counter in pharmacies.

The essential steps in drug development are shown in Table 1.

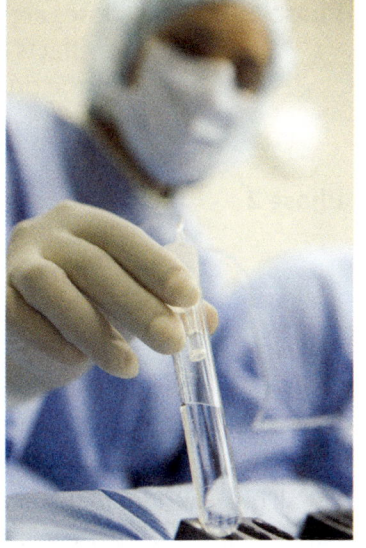

△ Fig 6 In vitro means in glass.

Step 1	New drugs come from one of two sources: either they are discovered in the natural world, or they are manufactured in the lab, sometimes by modifying existing drugs that are known to work, and sometimes by generating completely new classes of molecules.
Step 2	**In vitro** testing – **in vitro** means in glass. The effectiveness of the drug is tested on cells in culture, or on tissue samples. This works well when testing something like a new antibiotic, but is not much use for something like a painkiller or a sleeping tablet, which requires a whole, intact person who can communicate the results. In vitro testing can also provide useful information about whether a drug might be toxic to cells and tissues.
Step 3	Testing on animals. It is important to test the drug on whole organisms. Humans are mammals and so information gained from testing on other mammals is much more likely to produce useful information than testing on more distantly related species, such as insects. The drug needs to be tested for behavioural effects, some idea as to the correct dosage (known as dose–response studies), any side effects, toxic effects and, vitally, any **teratogenic** effects – can the drug cause malformed embryos and foetuses? The case of thalidomide may be over 50 years old but nobody is going to forget the lessons learned.
Step 4	If the drug makes it this far, it needs to be tested on real people. These are **clinical trials**.

△ Table 1

Apply this skill by doing Activity A1.

S10 CLINICAL TRIALS

This is the part of the drug development process where the drug is tested on patients. There are four standard phases, shown in Table 2.

Phase 1	Small doses are given to healthy volunteers to look for adverse effects on general health, metabolism and the way in which the drug is excreted from the body. These trials also help to identify the most suitable dose and best method of administration, for example, tablet or injection.
Phase 2	The drug is tested on patients who suffer from the specific condition. This allows the researchers to work out the relationship between dose and response. Obviously, the lowest effective dose needs to be established.
Phase 3	The main trial: a larger-scale trial where the effectiveness of the drug is compared to a placebo, or the current best drug. There may be multiple Phase 3 trials undertaken in different countries, and involving thousands of patients.
If it all goes well up to this point, and the drug is an improvement on existing drugs, a licence will be issued and medics can start prescribing the drug.	
Phase 4	All prescriptions are monitored for adverse side effects. Sometimes a drug will be withdrawn from the market but the process outlined above is so thorough that this is rare. Instead, a detailed profile of the most likely side effects will be compiled and issued with the drug.

△ Table 2

Apply this skill by doing Activity A1.

S11 SHARING THE KNOWLEDGE – SCIENTIFIC PAPERS, JOURNALS AND PEER REVIEW

When scientists have finished an investigation they write up their findings in research papers. These papers should contain:

- An **abstract** – a short paragraph that summarises the whole piece of research. This is very useful in helping potential readers to decide whether it's a paper that's going to interest them.
- An *introduction* that sets the scene and covers other relevant research.
- The *method*. This should include how the data was gathered, giving enough detail to allow others to repeat the investigation and reproduce the results, if they want.
- The results, in other words the **data** gathered, and a description of the way the data was processed and analysed.
- The **conclusion** – the interpretation of the results or what the results show, and what can be inferred from them.
- **References** – a list of all the sources used in the production of this piece of research. If you don't give credit for the work of others, that's **plagiarism**.

Journals are publications – think of them as academic magazines – that contain recent research papers about a particular area of science. The biggest breakthroughs in science will find their way into the most prestigious journals, such as *Nature*. The more obscure the work, the more obscure the journal, and the smaller the readership.

Most journals have such a small, specialised readership that they are never printed – they are just published online. There are tens of thousands of different journals being published to report new research all over the world.

Scientist	Journal	Peers	Journal
Scientist writes paper and sends it to journal	Paper checked by editor and sent to other scientists for peer review	Other scientists read paper and send feedback to journal	Based on feedback, editor decides whether or not to publish

△ Fig 7 The peer review process.

Scientists are not infallible. Sometimes the methods used in an investigation are not perfect, sometimes the interpretation of the results is open to debate and sometimes the research is done by people who are biased – such as drug companies saying that their own drugs are more effective than they really are, or not reporting some serious side effects. The vast majority of investigations are carried out completely genuinely, but there are many cases of scientists falsifying their results, usually to make a name for themselves or to make money.

So how do you make sure that the science is unbiased, rigorous and reliable – as good as possible? The answer is **peer review**. Peers, in this context, are scientists working in the same field as the writers of the paper. Scientific papers must contain all the methods and results so that other scientists, if they want, can repeat the investigation and get the same results. Peers are the people most likely to be able to spot problems.

So when scientists complete a piece of research, they write up their methods and findings in a scientific paper. This is then sent to a scientific journal for publication. Before publication, the editors of the journals send out the paper for peer review. Sometimes, the editors will require changes to be made as a result of the peers' comments. Work that is seen to be original and valid – in short, good science – will be published.

Apply this skill by doing Activity A11.

S12 ACCURACY, PRECISION, RELIABILITY AND VALIDITY

When **data** is collected, we need to make sure that it is suitable for its purpose. It needs to be collected using equipment of the right sensitivity, using the right methods and the right units.

Accuracy. Accurate data is close to the true value.

Precision. When describing data, the word precision has several different meanings. Generally, data is described as precise when the values are closely grouped together, without many anomalous results. However, precision can be subdivided into repeatability and reproducibility.

Repeatability. When the data is gathered by the same person, using the same equipment, at the same time.

Reproducibility. When the data is gathered by different people, using different equipment, at a different time.

△ Fig 8 This data gathering is possibly not as accurate as it could be.

The word "precision" can also be used to describe the equipment being used. An instrument that measures in microns is more precise than one that measures in millimetres but it might not necessarily be the right instrument for the job. For example, when measuring someone's height, using a metre ruler and saying "about 1.7 m" is not precise enough; 1.76 m is as precise as we need and 1.764927 m is clearly ridiculous. Once you have selected the right unit of measurement, you do not need more than two **decimal** places.

Reliability. If an investigation can be repeated by different people, using the same methods, and they get the same results, the results are reliable.

Validity. An investigation is valid if it is good science, meaning that the investigation is properly carried out, data is accurate and precise and, if the investigation were to be repeated, the same data would result and the conclusions were sound.

Apply this skill by doing Activity A11.

S13 RANDOM AND SYSTEMATIC ERRORS

Errors can occur when taking measurements, making them inaccurate.

Random errors can be above or below the actual value. This type of measurement is almost always present in a set of data for one reason or another. It may be **human error**. For example, if you measure your height to the nearest millimetre every day, you would get a range of different values, some above and some below your actual height. Random errors are usually difficult to eliminate. However, if you take a mean of all the readings, the random errors will tend to cancel themselves out.

Systematic errors occur when the readings are consistently above or below the true value. They are usually due to faulty equipment, or a flaw in the method – for example, a set of scales that always gives readings that are 8% higher than the true value. If you average out all the systematic errors, you will still get a value that is significantly different to the true value. Systematic errors are usually easier to eliminate than random errors, and correct calibration of equipment usually puts things right.

Apply this skill by doing Activity A15.

S14 DRAWING VALID CONCLUSIONS

Towards the end of every investigation comes a **conclusion**, which ideally should be one short, clear sentence, relating to the original hypothesis. Supporting the conclusion will usually be a statement along the lines of "The results were found to be significant at the (for example) 0.01 level." This means that we can be 99% certain that the results are significant (see S40 Statistical tests and values of p).

It's rare that a scientific paper will conclude "the results were not found to be significant" because if that were the case, there would be little point in publishing the paper. Of course, if your school or college investigations turn out not to be significant, then that must be your conclusion.

As an important part of peer review (see S11 Sharing the knowledge), other scientists will look at the following to evaluate whether the conclusions are valid or not:

- Are the measurements precise and free from anomalies?
- Are the measurements **accurate**?
- Are the measurements repeatable and **reproducible**?
- Is there any evidence of **bias**?
- Have any false assumptions been made?

The approach of trying to find fault with other people's work may seem to be aggressive or negative but it's neither; it's simply a very effective way of making sure that the science is rigorous, and is a solid foundation on which to build.

Apply this skill by doing Activity A11.

S15 HAZARDS, RISKS AND RISK ASSESSMENTS
When doing lab-based practical work

Teachers have a responsibility to discuss the risk assessment with the students. There is shared responsibility – students have to play their part too. You will already know the basic precautions: wear a lab coat, use safety glasses, tie long hair back, cover cuts and grazes with a sterile plaster, de-clutter the room so that desk tops and floor are clear.

You should be made aware of the following:

- The dangers of any chemicals being used, and what to do if they are spilt on the skin or if they get in the eyes.
- The dangers of using heat and glassware should be familiar.

- If you are dealing with bacteria, you should have been taught basic *sterile technique* so that the organisms you are working with do not get out into the environment, and so that organisms from the environment do not get in and contaminate your investigation.

When doing fieldwork

- Are you working near water? Water can be dangerous if it is deep, fast moving or dirty. If there is a danger of falling in and not being able to get out, you should not be doing fieldwork there.
- Infection risk. All habitats contain micro-organisms – if they were sterile, there would be nothing to study – and so will potentially contain pathogens. Sensible precautions include covering up cuts and grazes with plasters, and washing hands afterwards, especially before eating.
- Environmental considerations. Your work should minimise damage to the ecosystem you have studied. Rocks and logs should be placed back where they were found and there should be no litter left behind. Any organisms collected should be returned unharmed to the habitat.

△ Fig 9 You should always dress sensibly when in the lab.

Apply this skill by doing Activity A3.

S16 MODELS AND MODELLING

In biology, a model is anything that represents an idea, object, concept or process, so that it is easier to understand. Examples range from simple plastic representations of molecules and cells to very sophisticated computer programs.

Increasingly, computers can be used to model what will happen in a particular biological situation. As computers get more powerful we can model increasingly complex problems. Examples include:

- How fast will a particular virus spread?
- What will happen to the allele frequency when a particular selection pressure is applied?
- What will happen to the fish stocks if we continue to fish at our present rate?
- If we join amino acids in a particular sequence, what shape will the final protein be?

The genomes of most organisms are so big that they can only be effectively analysed using computers. The science of *bioinformatics* is an increasingly important area that relates to the storage and analysis of biological information by computers.

Apply this skill by doing Activity A10.

S17 WRITING FOR YOUR INTENDED AUDIENCE

A scientist who cannot communicate effectively isn't much of a scientist; at some point during your course you are going to be assessed on the quality of your written communication.

When doing any scientific writing, think about:

- Purpose – are you trying to deliver information, persuade people to think the same as you, or to entertain? In exams, you are delivering information.
- Audience – are they children or adults, scientifically trained or not? A good rule is: when in doubt, write for the intelligent general reader, which means you can write for an adult audience but you need to explain any specialist jargon. In exams, you can assume the examiner knows the jargon – you just need to use it correctly.
- Format – is it a scientific paper, exam short answer, exam essay, information leaflet, presentation or something different?

Apply this skill by doing Activities A1–15, AIS1, AIS2.

S18 ENSURING MEANING IS CLEAR

In examinations you will be awarded marks for clear and logical presentation and appropriate scientific language. At advanced level you might get away with the odd spelling mistake, such as *mitocondrion* (the correct spelling, by the way, is *mitochondrion*), but sentences such as "The liver is the most impotant organ in the body, and without it your dead" will lose you marks on several different levels. It's bad science – it's ludicrous to label an internal organ as *the most* important – and it's bad English. *Impotant* is a spelling mistake and using *your* instead of *you're* or *you are* is enough to make the examiners self-destruct.

In exam answers, beware of words like *it* or *they*. For example, in the question *"List two differences between mitosis and meiosis"*, an answer such as "It halves chromosome number" will gain no marks because it isn't clear which process you are referring to.

Apply this skill by doing Activities A1–15, AIS1, AIS2.

S19 ORGANISING INFORMATION CLEARLY AND COHERENTLY

Writing long answers

Long answer questions, usually for 4–8 marks, are found on most exam papers and though they may look scary they are often easy because they usually test factual recall; for example, "Outline the main features of the electron transport chain."

You need to match your answer to the marks available. If there are 6 marks available, read your answer back and ask, "Have I said six things relevant *that I learned since I started this course*?" Your answer should be in continuous prose but you probably won't lose more than 1 mark if you give your answer in bullet points (though this depends on your exam board).

Writing synoptic essays

Synoptic essays are set to assess whether you have an appreciation of biology as a whole. These essays will be set on 'big ideas' that draw information from different areas of the specification. With short-answer questions you just need to get to the point, but in longer answers you need to organise information clearly and coherently. Examples of synoptic essays set in recent years include *Cycles in Biology* and *The causes and significance of variation*.

As a general rule, you can write a page of A4 (standard exam paper size) in about 10 minutes, though it obviously takes longer if you have small writing. So if you have 35 minutes to produce the essay, you should spend about five minutes on the plan and then aim to write about three sides.

First, write a plan. Do not cross it out. Examiners can look at your plan and decide if you deserve extra marks, especially if you run out of time and don't finish the essay.

How do you go about planning a synoptic essay? There are two key elements.

i. You need a good working knowledge of what is on the specification, so you can trawl through it in your mind and pick the relevant sections. NB Don't forget plants! An alternative approach is to think about all of the different "levels" of biology:

- Molecules – cells – tissues – organs – whole organisms – populations – communities – ecosystems.

For example, in the *Cycles in Biology* essay, look at each level in turn:

- Molecules – ATP is constantly being recycled. Also the Krebs cycle and the Calvin cycle.
- Cells – mitosis and the cell cycle
- Whole organisms – the oestrus cycle and the cardiac cycle
- Ecosystems – the carbon and nitrogen cycle

ii. When revising, look at all the previous essays that have been set and do outline plans for them. Check them against the mark schemes that should be available online.

Apply this skill by doing Activities A1–15, AIS1, AIS2.

S20 USING SPECIALIST VOCABULARY

When writing long answers and essays, examiners expect your answer to reflect the fact that you have studied science for two years and have made some progress. You don't need to write long, convoluted sentences – stick to simple, clear sentences – but you do need to use words you have learned at advanced level.

For example, before you started your course you may have written something like: "*In the small intestine, starch is broken down into simple sugars.*"

An advanced version of the same sentence could be: "*In the ileum, the polysaccharide starch is hydrolysed first into maltose and then a second enzyme, maltase, hydrolyses the maltose into glucose.*" This answer shows that you have moved on to another level of detail and understanding. It also shows that you have improved your vocabulary and can use it appropriately. For example, the word *hydrolyse* shows that you know the process and the type of reaction involved.

Apply this skill by doing Activities A1–15, AIS1, AIS2.

S21 COMMAND WORDS IN EXAMS

Almost all exam questions have a command word that tells you what you need to do. You can match these command words to *Bloom's Taxonomy*. You may not have heard of *Bloom's Taxonomy* but you will probably be familiar with the following list of progressively more difficult and sophisticated thinking skills. Starting with the simplest, the basic six skills are:

1. **Remember** – simply learn the facts.
2. **Understand** – explain a particular idea or process. You can learn to draw and label a heart, for example, without having any idea about how it works.
3. **Apply** – once you have gained an understanding of basic principles and ideas, you can apply them to new situations.
4. **Analyse** – in science this usually means looking at data and deciding what it means. It also involves deciding what to do next.
5. **Evaluate** – weigh up, or look at both sides.
6. **Create** – in science, creative thought involves coming up with new hypotheses to ask, and new ways of investigating them.

Common command words in exams

Command words in exams generally follow *Bloom's Taxonomy*. Basically, the further down the list you go, the more likely you will be to perceive the question as a hard one. *Create* is so difficult that it rarely appears on exams.

Tables 3 and 4 show the command words used in International Baccalaureate Diploma Biology, but all the advanced exam boards are similar. Table 3 details what each command word is asking you to do, with example exam questions, and Table 4 shows how they relate to their classification in the taxonomy.

Command word	What it's asking you to do	Example of exam question
List	Give a series of names or points, but no more detail than that	List the main taxonomic groups in order from the most general to the most specific
Define	Give a precise meaning	Define the term 'community'
State	Give one precise piece of information	State Fick's Law
Draw	Use pencil lines to represent	Draw a standard population growth curve
Label	Add lines and words to show they you understand the diagram	Label the atrioventricular valves on the diagram of the heart
Distinguish	State the difference between two or more alternatives	Distinguish between the terms *allele* and *gene*
Describe	Tell us what you can see; explanation not needed	Describe the graph

(Table 3 continued)

Command word	What it's asking you to do	Example of exam question
Identify	Select an answer from the possibilities given	Identify the two most closely related species from the diagram
Calculate	Do some sums, and show your working	Calculate the growth rate shown by the curve
Estimate	Give a rough approximation; no need to calculate exactly	Use the graph to estimate the water potential of the cytoplasm
Annotate	Write little notes all over the diagram to show your understanding	Annotate the trace of the action potential with details of ion movement
Outline	Give us the main points, but don't go into too much detail; it is vital to match your answer to the marks available	Outline the essential feature of the electron transport system
Suggest	This won't be something you have learned; you are being asked to give an answer based on the information given – any sensible answer will be acceptable	Suggest an alternative conclusion for these results
Predict	Tell us what you think will happen	Use the graph to predict what the population will be in 2020
Evaluate	Weigh up; look at both sides of the argument; talk about the implications and limitations	Evaluate the conclusion "alcohol is good for your health"
Sketch	Draw a diagram but don't worry about the details; if it's a graph, the basic shape is important but don't worry about values on the axes	Sketch a graph of the effect of substrate concentration on enzyme activity
Construct	Design/build – usually a results table or graph; you need to decide which format is most suitable	Construct a results table in which to record your data
Deduce	Figure it out – reach a conclusion from the information given	Deduce the genotype of the parents from the offspring's genotype
Show	More common in maths exams, it means "explain and include all the steps"	Show that this particular gene is not sex linked
Explain	Give reasons for your answer; show that you understand the reasoning or causes; use words like *so* and *because*	Cone cells allow us to see in colour; explain how

(Table 3 continued)

Command word	What it's asking you to do	Example of exam question
Determine	Find the right answer, usually from a range of possibilities	Determine which type of cell is responsible for high visual acuity
Compare	Tell us about the similarities and differences	Compare the processes of meiosis and mitosis
Discuss	Similar to evaluate – you need to give both sides of an argument	Discuss the issue of hormone replacement therapy

△ Table 3

Objective	Examples of command words	Bloom's Taxonomy
Demonstrate understanding of facts, concepts, methods and techniques.	List, define, state, draw, label	Remember, understand
Apply and use facts, concepts, methods and techniques.	Distinguish, describe, identify, calculate, estimate, annotate, outline, suggest, explain	Understand, apply
Construct, analyse and evaluate questions about the scientific method.	Suggest, predict, evaluate, assess, sketch, deduce, show, comment, explain, calculate, determine, compare, derive, design, solve	Analyse, evaluate

△ Table 4

Apply this skill by doing Activities A1–15, AIS1, AIS2.

S22 UNITS OF SIZE IN BIOLOGY

The standard unit of length is the metre (m), but most of the measurements taken by biologists use smaller units. Generally, there are only three units you will ever need:

- a *millimetre*: one thousandth of a metre
- a *micrometre*: one thousandth of a millimetre, commonly called a micron and written as μm (μ is the Greek letter *mu*)
- a *nanometre*: one thousandth of a micrometer. Commonly written as nm.

Apply this skill by doing Activity A4.

S23 BIOLOGICAL UNITS AND STANDARD FORM

Standard form is a simple way to deal with large or small numbers. It always takes the form of $A \times 10^n$ where A is a number between 1 and 10. For example, 8 million is written as 8 000 000 or, in standard form, 8×10^6.

For most International System (SI) units, you need to be able to scale up or down in thousands – powers of 10^3. For units of size:

- a millimetre is written as 10^{-3} m
- so a micrometre is 10^{-6} m
- and a nanometre is 10^{-9} m.

Apply this skill by doing Activities A3, A5.

S24 UNITS OF VOLUME AND WEIGHT

The **SI unit** of volume is the litre, which can be written as a lowercase letter l ("el") However, this may be confused with the number 1, so it is better to write it in full. To convert to old units, it's useful to remember "a litre of water is a pint and three quarter". A litre is actually 1.76 pints, but that doesn't rhyme.

In a similar way to size, a lot of work in biology involves small volumes.

A litre is commonly written as dm³, which means decimetre cubed.

A decimetre is 10 cm, so a decimetre cubed is 10 × 10 × 10, which is 1000 cubic centimetres, or 1000 cm³.

1 cm³ is also called a millilitre, often abbreviated to ml, pronounced *mil*.

So, just as with units of size, we can get smaller volumes by factors of one thousand, or powers of 10^3:

- a millilitre is 10^{-3} litre
- so a microlitre is 10^{-6} litre – much smaller than the average raindrop but commonly used in genetic engineering investigations. A micropipette would be used to measure out volumes this small.
- and a nanolitre is 10^{-9} litre, which is rarely used in investigations because it is impractical to deal with volumes this small.

The **SI units** of size, volume and weight are connected by the properties of water, which makes sense because it is a pretty common substance.

The original definition of a litre is "the volume occupied by 1 kilogram of water at its maximum density". One litre of water weighs 1 kilogram, and so 1 cm³ of water weighs 1 gram (1 g).

Apply this skill by doing Activities A1, A5.

S25 MICROSCOPES AND MAGNIFICATION

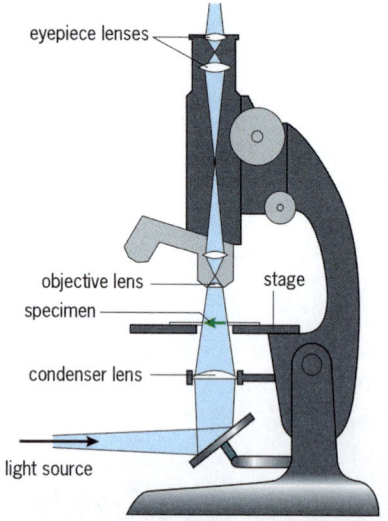

Δ Fig 10 (a) The light microscope uses two lenses. The total magnification can be worked out by multiplying the eyepiece lens (often × 10) by the objective lens (usually × 10, × 25 and × 40).

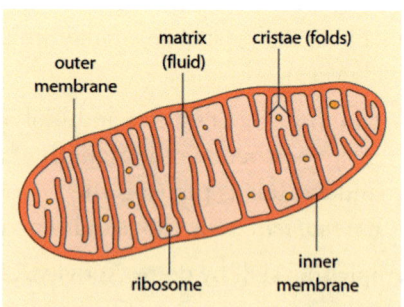

Δ Fig 10 (b) A diagram of a mitochondrion drawn from a micrograph. When you are given a scalebar like this, you can use it to measure the magnification. You'll need to measure it with your ruler. This one was 15 mm on the printed page, but it represents one µm. 15 mm is 15 000 µm so this scalebar tells you that the image had been enlarged ×15 000.

There are two basic types of microscope:
- *light microscopes*, which use light, focused by lenses
- *electron microscopes*, which use beams of electrons, focused by magnets.

The magnifying power of the electron microscope is much greater than that of the light microscope, but the ability to magnify is no use without an improved **resolution**, which is the level of detail that can be seen. For example, if a microscope has a resolving power of 10 nm, any objects closer together than 10 nm will appear as one object.

The power of the light microscope is limited by the wavelength of light. The wavelength of a beam of electrons is a lot smaller, and the resolution is correspondingly higher.

When using microscopes, and the images they produce, there are three important values:

i. The *actual size* – the size it is in real life.
ii. The *magnification* – ×100 means that the image is 100 times larger than it is in real life. Some electron microscopes can enlarge 500 000 times. At that sort of magnification, a printed full stop becomes larger than a football pitch.
iii. The *observed size* – the size of the photograph or diagram. You'll have to get your ruler out.

Almost all exam questions will give you one value and expect you to get the second one by measuring. Once you've got two values, working out the third one is easy because

$$\text{Actual size} = \frac{\text{observed size}}{\text{magnification}}$$

This is often abbreviated to $A = \frac{O}{M}$

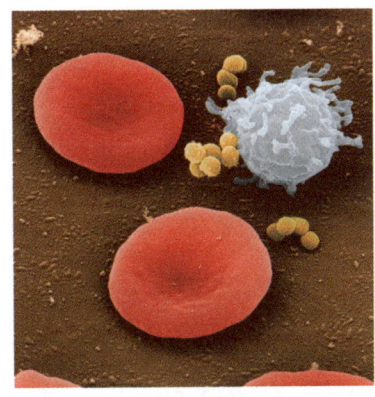

△ Fig 10 (c) Micrographs are photos taken with microscopes. This one shows red and white blood cells surrounded by some bacteria (orange), taken with a scanning electron microscope. The colour is not natural – it is added by the computer.

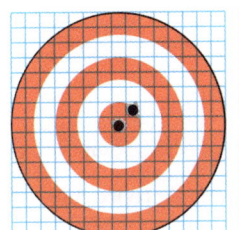 

higher resolution – the instrument can detect the difference in position between the two shots

lower resolution – the instrument cannot detect the difference between the two shots

△ Fig 11 Resolution is the ability to see detail.

You need to be able to rearrange the formula to make O or M the subject.

So, *observed size* = actual size × magnification

And *magnification* = $\frac{\text{observed size}}{\text{actual size}}$

Apply this skill by doing Activity A4.

S26 PERCENTAGES AND ESTIMATES

Being asked to work out a percentage is probably the commonest mathematical question in biology exams. The term *per cent* means per hundred. They are very useful values because they allow comparisons. For example, the statistic "in the USA, 30% of adults are obese" allows a meaningful comparison with other countries with different populations.

In the exam situation it is easy for your mind to go blank and select the wrong formula. A good tactic is to think about what 1 per cent, or 10 per cent, is. You should then be able to make a reasonable estimate as to what the answer should be before you do the calculation. If the calculation then gives a silly answer, try it again.

The basic formula is this:

% change = $\dfrac{\text{difference between original and new value} \times 100}{\text{Original value}}$

For example:

In the year 2000 the number of cases of asthma in a city was 4400. By 2012 it had reduced to 3500. Calculate the percentage decrease.

Think about it. The original figure was 4400, and it has decreased by 900. One per cent of 4400 is 44 and 10% is 440. So 900 will represent a decrease of just over 20%. With that estimate in mind, you can do the calculation with confidence.

The calculation you should perform is

$\dfrac{4400 - 3500}{4400} \times 100$

$= \dfrac{900}{4400} \times 100$

$= 0.2045 \times 100 = 20.45\%$

If you got an answer of 488%, you got the formula upside down.

Apply this skill by doing Activity A1.

S27 RATIOS

Ratios are comparisons between two values that measure the same thing, such as the weight of a potato cube before and after soaking in water.

There are two ways of writing ratios. Either:

- the familiar way with a colon in the middle, such as 3:1, where the colon means *to*. In biology, this type of ratio is often used in genetics. For example, 3 long haired:1 short haired

or:

- as a single number, sometime called a quotient, which is a product of dividing the number on the left by the number on the right, so that 1:4 becomes 0.25.

An example of how ratios are commonly used comes in the common osmosis practical that uses pieces of potato left to soak in a range of concentrations of salt. The chips are weighed before and after soaking. The final result is expressed as a ratio of final weight:initial weight. If the ratio is more than 1, the potato increased in size. If it is less than 1, it shrank. If the ratio was exactly 1, it stayed the same.

Apply this skill by doing Activity AIS2.

S28 MEAN, MEDIAN AND MODE

For any given measurement, it's very useful to have an idea of a typical value. If we say that the average adult cat weighs 4.5 kg, everybody knows that there is a range of values; some cats are heavier and some are lighter. The term average means the middle. However, in science there are three different types of average: mean, median and mode, which is why examiners don't like you using the word average when referring to the mean.

For example, blackbirds have two to three broods a year, each containing three to five chicks. So you might study a number of pairs over one breeding season and count the number of chicks raised. The results might look something like this:

12, 7, 9, 12, 8, 11, 11, 9, 13, 12, 12, 15, 9

△ Fig 12 A brood of blackbird chicks.

Apply this skill by doing Activity A15.

Mean

The arithmetic mean is what most people think of it as *the average*. It is obtained by adding up all the values and dividing by the number of samples, which in this case works out as 10.77. Means are useful when all the values are relatively close together, and there are no extreme values. However, if there was a particularly unlucky pair of blackbirds that managed to raise only one chick all year, it would **skew** the mean, giving a lower, less accurate mean value.

Median

The median is the middle value. You have to rank the values in order and select the middle one.

7, 8, 9, 9, 9, 11, 11, 12, 12, 12, 12, 13, 15

There are 13 values and so the seventh one is the median – there are six either side – so in this case it is 11. The advantage of using the median is that extreme values don't affect the results. Medians are especially useful when dealing with **ordinal data** – data that is in rank order – or with frequency data. An example of frequency data would be a list of all the girls' shoe sizes in a school or college.

Mode

The mode is the commonest value. Looking at our blackbirds, the mode is 12. Modes are the only measure of average that can be used with **nominal data**, which is data that has names rather than numerical values. For example, if you asked a group of students which football team they support, the mode would be the best-supported team.

Apply this skill by doing Activity A15.

S29 PERCENTILES, DECILES AND QUARTILES

Percentiles are used to compare a particular value with all other values in the range. A common example of percentile is the measurement of a child's progress, such as height and weight. If a five-year-old weighs 25 kg, the graph will show that they are on the 95th percentile, which means that 95% of five-year-old children are below that weight, and only 5% are above it.

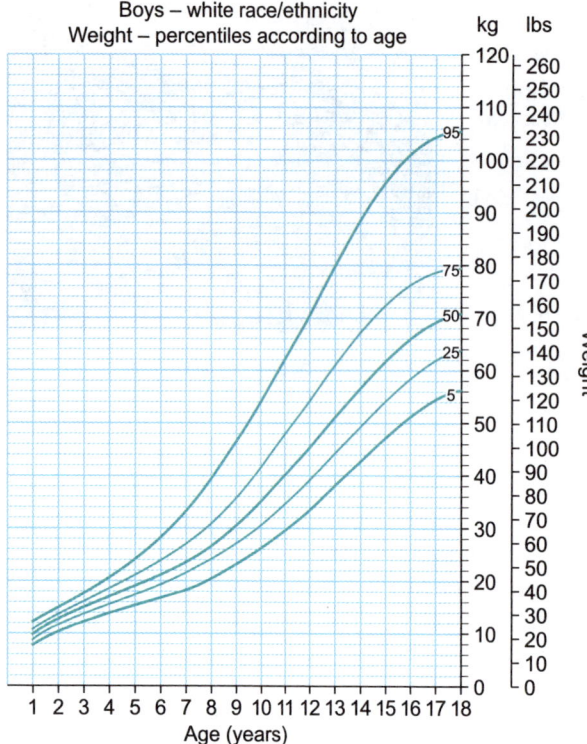

△ Fig 13 Child's weight chart, measured against age.

Deciles and **quartiles** are similar measurements to **percentiles**, they just involve larger sections of the graph. Deciles are increments of 10% while quartiles (quarters) are increments of 25%. The interquartile range is the middle 50% of a range of values.

Apply this skill by doing Activity A15.

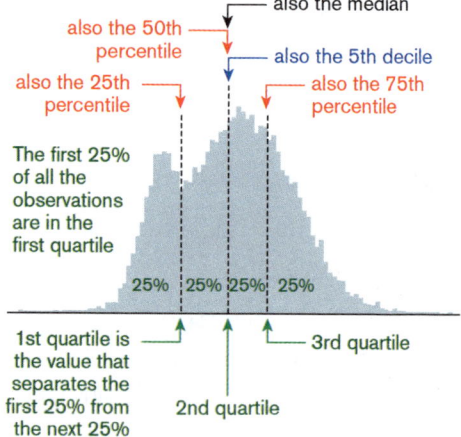

△ Fig 14 Quartiles, deciles and percentiles.

S30 WORKING WITH CONCENTRATIONS

The word concentration refers to the amount of dissolved substance – called **solute** – per unit volume of water. In biology the solvent is always water.

There are two commonly used methods of measuring concentration: molarity and percentage.

Molarity. You may remember from basic chemistry that a mole of a substance depends on its molecular mass. For example, the molecular mass of glucose is 180.16 and so one mole of glucose weighs 180.16 g. To make a molar solution of glucose you need to weigh out 180.16 g of glucose, tip it into a measuring cylinder and then top it up to 1 litre; this is a 1.0 M solution of glucose. *NB* You do not add the glucose to 1 litre of water.

In biology, molar solutions are usually a bit strong and so you need to be able to make dilutions. For example, if you were given 100 cm^3 of 1.0 M glucose you could dilute it by adding water. For a 0.1 M solution you would take 10 cm^3 of the 1.0 M solution and add 90 cm^3 water.

If you were given a 1.0 M solution you could make a range of dilutions, as shown in Table 5.

Concentration	1.0	0.8	0.6	0.4	0.2	0.0
Volume of 1.0 M glucose	20	16	12	8	4	0
Volume of water	0	4	8	12	16	20

△ Table 5

Percentage solutions. These are quick and convenient because there is no need for careful measurement of weight; they just use volumes. To make, say, a 5% solution of glucose you would pour 5 g of 1.0 M glucose into a measuring cylinder and top it up to 100 cm with water. If you just need a rough approximation, this is a quick and easy method.

Serial dilutions

It's easy to make a big range of concentrations. In six easy steps you can make a solution that is one millionth of your original solution.

1. Take 1 cm³ of your original 1.0 M solution and add 9 cm³ of distilled water. You now have a 10% solution, or 10^{-1} mole.
2. Take 1 cm³ of the 10% solution and dilute that in 9 cm³ of distilled water. You now have a 1% solution, or 10^{-2} mole.
3. Repeat this simple process four more times to produce solutions of 10^{-3}, 10^{-4}, 10^{-5} and 10^{-6}.

Apply this skill by doing Activities A3, AIS2.

S31 WATER POTENTIAL

Osmosis is a simple idea; solute attracts water. Particles such as sugar, sodium and chloride dissolve because they attract and become surrounded by a layer of water molecules. These water molecules are therefore not free to bounce around the way they would in pure water.

All solutions and cells have a particular water potential, which is a value that is given the symbol Ψ – the Greek letter psi, pronounced *sigh*. The value of water potential is expressed in kPa (kilopascals), which are units of pressure. The water potential is a measure of the tendency of the solution or cell to absorb water by osmosis.

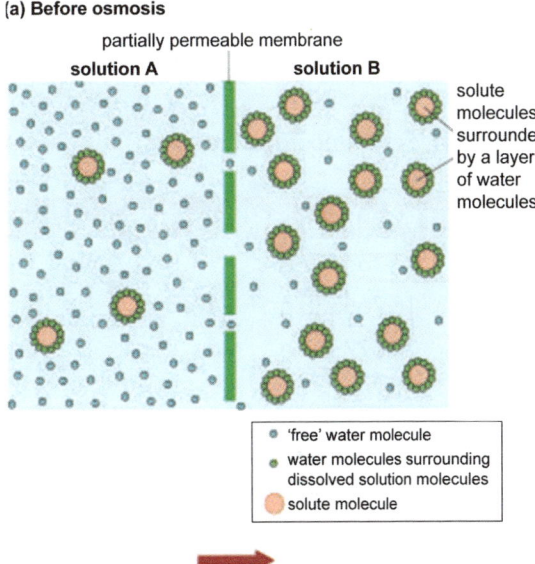

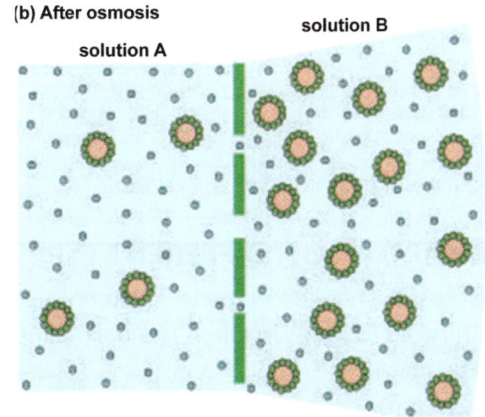

△ Fig 15 To understand osmosis, you need to understand what happens when a substance dissolves in water. The dissolved substance becomes surrounded by a shell of water molecules which are not free to move around. The more solute there is, the more water molecules it "ties up".

Water potential is a negative scale. Pure water has the highest possible water potential: 0. The more dissolved particles, the lower the water potential. Next time you put sugar in your tea, think: "I'm lowering the water potential." Apply this skill by doing Activity AIS2.

S32 HOW TO CONSTRUCT A RESULTS TABLE

One of the key skills you need to develop at advanced level is the ability to design your own results table and the only way to do this is to practise.

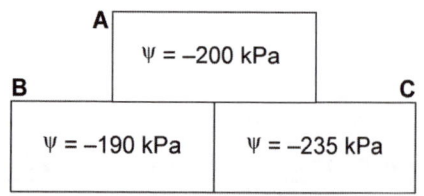

△ Fig 16 A common exam question – "Predict the direction of net water movement between these three cells." Water will diffuse from the areas of high water potential (Ψ) to the areas of low water potential. So water will pass from A and B into C, and also from B to A.

The golden rules for making a results table are:

- Clearly label the units at the top of columns and rows.
- Do not repeat the units in the individual boxes.
- Use an appropriate number of **decimal** places: 2 is usually best.
- If you have taken repeated readings, you should also include the mean of those readings.

Table 6 gives us a poor results table.

Temp	Weight of seedlings group A	Weight of seedlings group B
20 °C	15.363477 g	14.33556 g
30 °C	18.766588 g	14.4455533 g
40 °C	20.677335 g	12.9967 g

△ Table 6

Table 7 gives a much better one, showing the same data.

Temp (°C)	Mean weight of seedlings (g)	
	Group A	Group B
20	15.36	14.34
30	18.77	14.45
40	20.68	13.00

△ Table 7

Apply this skill by doing Activity AIS2.

S33 WHAT TO PLOT? DIFFERENT TYPES OF GRAPH

Graphs exist to show data clearly and simply and, if possible, the relationship between two different variables. The best graph to draw depends on the type of data you have.

Qualitative data does not have a numerical value. Examples would be eye colour, blood group, type of cancer. This is sometimes called categorical or discontinuous data.

Quantitative data has a numerical value. Examples include time, temperature, pH, speed/rate, height and weight. This is sometimes called continuous data, because each value will be a reading somewhere along a continuous scale.

Table 8 shows when you need to use the different charts, graphs and diagrams.

Type of graph	Diagram of graph	Independent variable: x axis	Dependent variable: y axis	Used	Example
Bar chart		Qualitative: separate groups	Quantitative: often "how many individuals"	To compare difference between groups or categories	Numbers of new cases of different types of cancer
Histogram		Quantitative, continuous data	Quantitative: often "how many individuals"	Shows distribution: e.g. exam marks, height	Number of specimens in each size range
Line graph		Quantitative, continuous: e.g. time	Quantitative, continuous	Shows how two continuous variables are related	How pH affects enzyme activity
Scatter graph		Quantitative, continuous	Quantitative, continuous	Looks for degree of correlation	Blood cholesterol levels vs incidence of heart disease
Pie chart		Shows the contribution of various factors to a whole			Causes of death
Venn diagram		Shows similarities and difference, or degree of overlap, between two areas			Compare anatomical features of reptiles and birds

NB The difference between a histogram and a bar chart is that histograms have equally spaced ranges of values, such as 1–5, 6–10, 11–15, and the bars touch. In bar charts, the categories are not related and the bars do not touch.

△ Table 8

Apply this skill by doing Activity AIS1.

S34 HOW TO PLOT A LINE GRAPH

Step 1 Decide which axis is which. The independent variable (cause) goes on the *x* axis (remember "X is a cross") and the dependent variable (effect) goes on the *y* axis.

Step 2 Work out a scale. You do not have to start at zero. Look at your highest and lowest values and match it up to the graph paper you have available. The scale should use up at least half of the width and height of the paper.

Step 3 Plot the points. Use a sharp pencil so they are as accurate as possible.

Step 4 There is no single *right* way to join the points. Teachers and examiners are constantly changing their minds. You can join the points with a ruler, join the points freehand or estimate a **line of best fit**. Alternatively, if you do the whole thing on a computer it will draw the line for you. *NB* In biology, if you are asked, "Why did you join the points with a straight line?" the answer is, "Because there were no readings in between."

Apply this skill by doing Activities AIS1, AIS2.

S35 READING GRAPHS

A **line graph** can be used to predict values for which no actual readings were taken.

Interpolation is the name given to the prediction of values that occurs between actual measured values. In the population graph shown, no reading was taken for 1960 but we can interpolate the value using the line (it's about 3 billion).

Extrapolation is the prediction of values by carrying the line on beyond the value taken. This graph extrapolates world population growth until 2050. However, there are a lot of assumptions made when predicting that far into the future.

Apply this skill by doing Activities AIS1, AIS2.

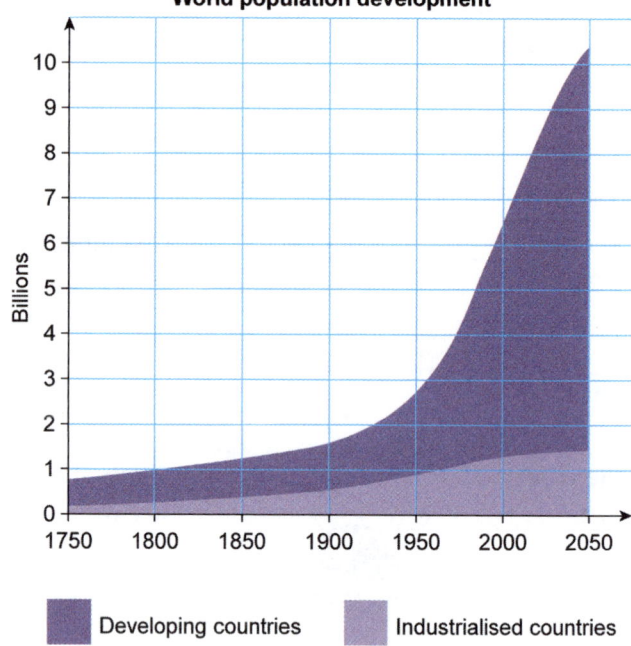

△ Fig 17 You may be able to predict values from a line graph.

S36 CORRELATIONS AND SCATTERGRAMS

A **scattergram**, scatter graph or scatter diagram – they're all the same – is a graph on which two sets of data can be plotted so see if there is a **correlation**. For example, you could plot the levels of cholesterol in the blood and see if there is a correlation with the incidence of coronary heart disease.

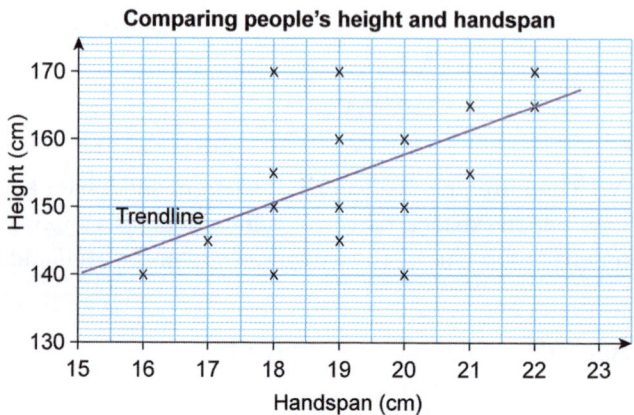

△ Fig 18 Scattergram comparing people's height and handspan measurements.

Computers can be used to draw lines of best fit through the data. There is a good correlation if most of the points are on or close to the line. Remember that a correlation is not the same as a cause.

Apply this skill by doing Activity A15.

S37 LOGARITHMIC SCALES

A logarithm is an index, like the 12 in 10^{12}. **Logarithmic scales** – or just log scales – can be very useful when you have to plot a wide range of numbers on one axis.

For example, a student needs to plot the following numbers on the x axis: 6, 12, 25, 50, 1050, 15 000, 300 000. It's impossible to get those numbers onto one scale with any accuracy. If your scale went from 0 to 300 000 on a normal-sized piece of paper, the first four values would be on top of each other and indistinguishable. The solution is to use a log scale in which the increments are 10, 10^2, 10^3, 10^4 and 10^5. You can get logarithmic graph paper.

Apply this skill by doing Activity A10.

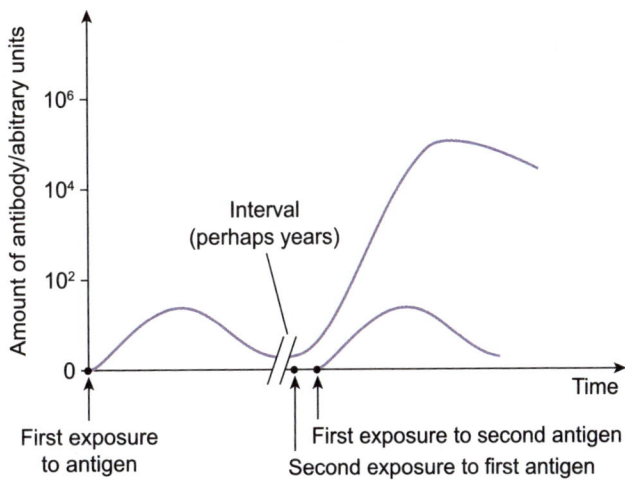

△ Fig 19 During the immune response, antibody production increases from a few molecules per unit of blood to a hundred thousand times as much. To fit this all on one scale, a log scale is needed in the y axis.

S38 SAMPLING

The idea of **sampling** is that you can get representative data without measuring every individual. Do you have to measure the height of every tree in a forest to get a fair idea of the mean height? Here are two commonly used techniques for deciding when you have got enough data.

i. When taking a particular measurement, use a **running mean**. Every time you take a measurement, calculate the mean of all the values taken so far. When the mean stops changing by more than a fraction of a per cent, you can stop sampling. Taking more readings will simply waste time.

ii. When studying diversity – the numbers of different species in a given area – how many **quadrats** do you have to throw? In a similar way to the running mean, you record the number of new species found at each quadrat. When the graph levels off, and you are finding no new species, you have done enough quadrats.

Apply this skill by doing Activity A14.

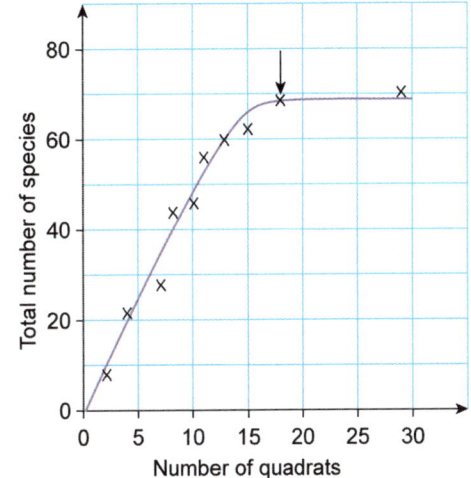

△ Fig 20 Defining the sample and collecting data are essential parts of empirical research.

S39 HARDY-WEINBERG EQUILIBRIUM

This is basically a clever bit of maths that is used in population genetics. There are two important applications of the Hardy-Weinberg equilibrium:

i. It allows you to calculate the frequency of genotypes in a population. Given the frequency of the homozygous recessive genotype, which is easy because you can count the individuals, you can work out how many are homozygous dominant and how many are heterozygous.

ii. It can be used to assess whether allele frequencies are changing from one generation to the next. This is important because you can tell whether a species or population is evolving or not.

△ Fig 21 The remarkable camouflage of the speckled footman moth.

Imagine there is a population of 100 moths. Their wing colour is controlled by one gene with two alleles. Allele B codes for the normal speckled colour while allele b codes for a black colouration. It's easy to count the number of moths that have genotype bb because we can see them. They're black. The magic of the Hardy-Weinberg equilibrium is that we can work out how many of the speckled moths are BB and how many are Bb – who all look the same – just by counting the black moths.

Here's how it works.

If p stands for the frequency of the B allele, and q stands for the frequency of the b allele, then

$$p + q = 1$$

This simply means that if you add up all the B alleles and all the b alleles, they must add up to 1, or 100%, because there are no other alternatives.

Each organism has two copies of this allele, and so their genotype can be BB, Bb or bb. This gives rise to the equation

$$P^2 + 2pq + q^2 = 1$$

- In this equation, P^2 is the frequency of the BB genotype.
- $2pq$ is the frequency of the Bb genotype, because there are two ways of generating that genotype (Bb or bB).
- And q^2 is the frequency of the bb genotype.

So this second equation is simply saying that if you add up all the individuals that are BB, Bb and bb, you get the whole population, 1.

The Hardy-Weinberg rules states that the allele frequency *will not change* from one generation to the next *unless*:

- There is migration in to or out of the population. This causes gene flow.
- Mating is non-random: if one genotype is more likely to mate with another particular genotype. For example, if the speckled moths seek out other speckled moths and avoid the black ones.
- The gene mutates to the point where there are more than two alleles.

- The population is small, in which case *chance* plays a large part in determining allele frequency.
- One genotype gives an individual a *selective advantage*. In this case, that genotype will reproduce more successfully than others. This is natural selection.

When allele frequencies change from one generation to the next, and this is due to the last reason and not the first four, that is evolution in action.

Apply this skill by doing Activities A8, A9.

S40 STATISTICAL TESTS AND VALUES OF p

With any investigation, you want to know if the results are significant or not. Statistical tests are used to help you decide. The end product of most statistical tests will be a value of p: probability. This will tell you the probability that your results are due to chance.

If you decide that your results are significant – and not simply due to chance – you can reject the null hypothesis. In doing so, you accept your experimental hypothesis. You have not *proved* anything, but you have gathered support for your hypothesis.

The value of p will tell you the probability that your results are due to chance, and therefore what your conclusion must be. p can only have a value between 0 and 1. The lower the value of p, the better. A value of 0.5 means that you can be 50% certain that your results are significant, which is clearly not good enough. So how certain do you need to be?

Scientists have decided that, in most circumstances, a reasonable threshold is 95%. If you can be 95% certain that your results are significant, you can reject the null hypothesis and say that your results are significant.

Converting these percentages into **decimals**, the vital value of p is 0.05, which means 5%. If p is equal to or less than 5%, we can be 95% sure that our results are significant. In investigations, you may see this written as $p \leq 0.05$, meaning that p is less than or equal to 0.05, or 5%. If results are even more significant, you might see a value like $p \leq 0.01$, meaning that we can be more than 99% certain that these results are not due to chance.

NB In the scientific world, almost all statistical tests are done on a computer. Spreadsheets such as Excel will do most straightforward statistical tests for you. The days of scientists doing hours and hours of longhand statistics are long gone.

Apply this skill by doing Activity AIS3.

S41 CHOOSING THE RIGHT STATISTICAL TEST

The flowchart (Fig 22) should allow you to choose a suitable statistical test depending on the type of data you have gathered. There are many more statistical tests available, but these are the basic ones that are suitable for advanced level study.

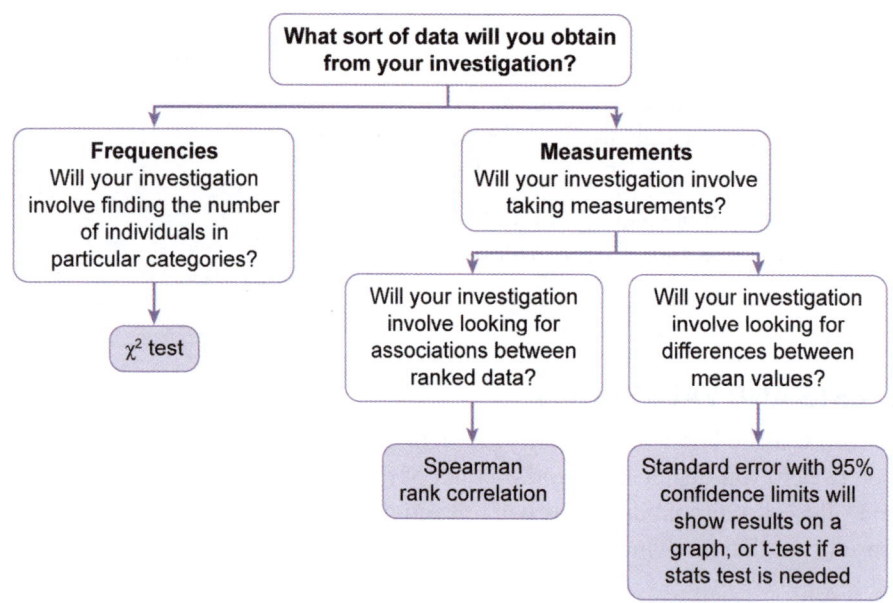

△ Fig 22 Choosing a suitable statistical test.

Apply this skill by doing Activity AIS3.

S42 NORMAL DISTRIBUTION AND STANDARD DEVIATION

Many biological measurements show **continuous variation**: a range of values with most in the middle and fewer at the extremes. There are countless examples: human height, weight and intelligence, exam results, the number of leaves on a tree. When values are shown on a graph, a bell-shaped curve known as a **normal distribution** is seen.

It's useful to know the spread of the data about the mean. This tells you the range in which most values fall, giving an accurate impression of what constitutes a normal value. By definition, one **standard deviation** contains the middle 68% of values, which is 34% below the mean and 34% above. Two standard deviations included the middle 95% of values.

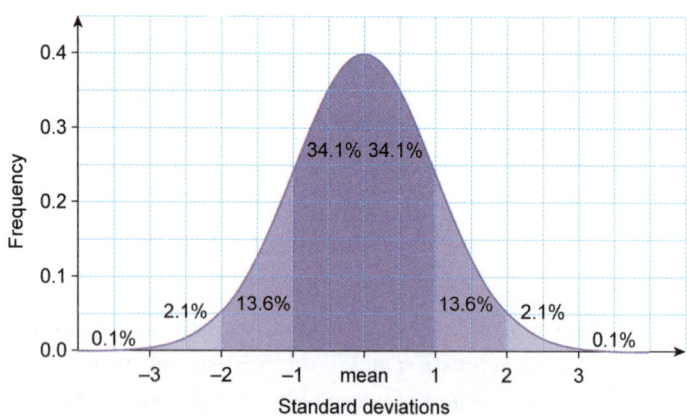

△ Fig 23 Normal distribution with standard deviations.

As an example, the standard deviation of human height would be written as 1.75 m ± 5 cm, meaning that the mean height is 1.75 m and 68% of the population fall between 1.70 m and 1.80 m. Standard deviations like this can be shown as **error bars** on the graph. If the error bars of two values do not overlap, you can say they are significantly different.

Apply this skill by doing Activity A11.

S43 STANDARD ERROR WITH 95% CONFIDENCE LIMITS

This is a test to show, on a graph, if there is a significant difference between two sets of values. For example, you might be testing a new fertiliser and you want to know if the new one is significantly better than the old one.

You would take all the plants in one batch and weigh them, and then work out the mean. You would then work out the **standard deviation** and then plot **error bars** which are 2 standard deviations above and below the mean. If the error bars do not overlap, you can be 95% certain that the values are significantly different.

Apply this skill by doing Activity A11.

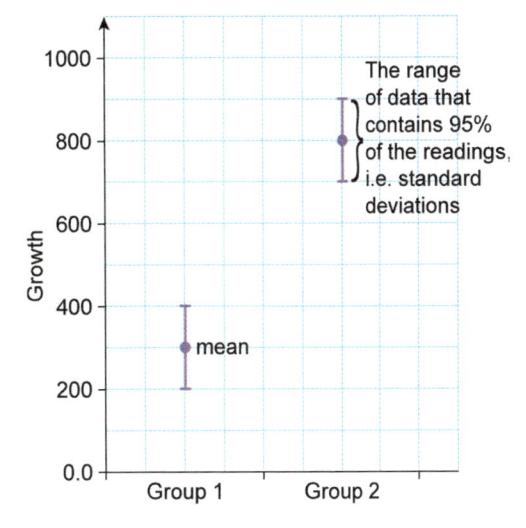

△ Fig 24 Graph showing means and error bars.

S44 THE CHI-SQUARED TEST

Chi-squared is used to decide whether there is an association between different categories, by comparing the observed results and those you would expect if they were due to chance. For example, you might have the hypothesis "the red squirrel prefers to nest in a particular species of tree".

You would then record the species of tree in which red squirrels build their nests. The results could be as in Table 9.

Species of tree	Number of squirrel nests observed	Number you would expect by chance (70/4)
Larch	12	17.5
Spruce	18	17.5
Scots pine	15	17.5
Fir	25	17.5
Total nests	70	70

△ Table 9

It's obvious that there are different numbers in each category, but are they *significantly* different or just due to chance? The chi-squared test can help you decide.

The formula for chi-squared is

$$\chi^2 = \frac{\Sigma(O-E)^2}{E}$$

where O is the observed results, E is the results we would expect by chance and Σ means the sum of all.

Apply this skill by doing Activity AIS3.

S45 STATISTICS – THE t-TEST

The **t-test** is a simple, common test used to find out whether there is a significant difference between two sets of data. t-tests work by comparing the difference between the means and the standard errors of two sets of data. The bigger the difference in the means, and the smaller the standard errors, the more likely it is that the groups are significantly different.

Apply this skill by doing Activity A11.

S46 STATISTICS – THE SPEARMAN RANK TEST

Spearman rank is a statistical test used to measure the degree of similarity or dissimilarity between two variables. For example, if you had the hypothesis "the smaller the seed, the further it falls from the tree", you would gather two sets of data – the weight of the seed and the distance it fell from the tree. You would then rank the data: lightest seed to heaviest, and greatest distance to smallest. If your hypothesis is correct, the two lists should match up. Lightest goes furthest, second lightest goes second furthest, heaviest goes least far. The Spearman rank test will give you information about the strength and direction of the correlation between your two sets of data.

The end product of the Spearman rank test is a number. A value of 1 indicates a perfect positive **correlation**. For example:

1 2 3 4 5 6 7 8 9 10 matches perfectly with

1 2 3 4 5 6 7 8 9 10. That would give a correlation of +1.

A score of –1 indicates a perfect negative **correlation**. In this case,

1 2 3 4 5 6 7 8 9 10 would have a perfect negative correlation with

10 9 8 7 6 5 4 3 2 1.

But do the sequences need to match perfectly in order to say that your results are significant? Well, no; the sequence

1 2 3 4 5 6 7 8 9 10 would still be significantly correlated with

1 2 3 5 6 4 7 8 9 10.

Basically, the more pairs you have, the more leeway there is to have some values out of sequence.

Apply this skill by doing Activity A7.

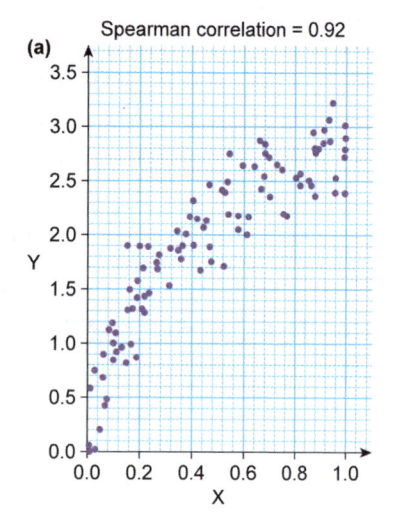

△ Fig 25 (a) shows good positive correlation.

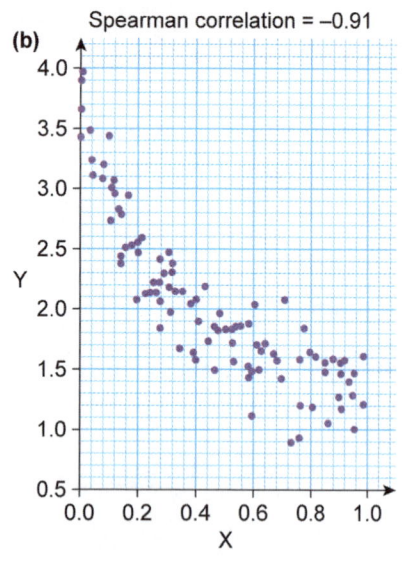

△ Fig 25 (b) shows good negative correlation.

Skills to Activities table

This table lists the activities that practise each skill.

Skill type	Skill	Links to activities
Working Scientifically	1 Science today	A4
	2 Theories, ideas and hypotheses	A4
	3 The scientific method	A1, A2, AIS1, AIS2, AIS3
	4 Testable ideas	AIS3, A11
	5 Control experiments	A2, A11, AIS1
	6 Investigations involving people	A1, A13
	7 The placebo effect	A1
	8 Blind, double-blind and open label trials	A1
	9 Drug development	A1
	10 Clinical trials	A1
	11 Sharing the knowledge – scientific papers, journals and peer review	A11
	12 Accuracy, precision, reliability and validity	A11
	13 Random and systematic errors	A15
	14 Drawing valid conclusions	A11
	15 Hazards, risks and risk assessments	A3
	16 Models and modelling	A10
Quality of Written Communication	17 Writing for your intended audience	A1–15, AIS1, AIS2
	18 Ensuring meaning is clear	A1–15, AIS1, AIS2
	19 Organising information clearly and coherently	A1–15, AIS1, AIS2
	20 Using specialist vocabulary	A1–15, AIS1, AIS2
	21 Command words in exams	A1–15, AIS1, AIS2
Maths	22 Units of size in biology	A4
	23 Biological units and standard form	A3, A5
	24 Units of volume and weight	A1, A5
	25 Microscopes and magnification	A4
	26 Percentages and estimates	A1
	27 Ratios	AIS2
	28 Mean, median and mode	A15
	29 Percentiles, deciles and quartiles	A15
	30 Working with concentrations	A3, AIS2
	31 Water potential	AIS2
	32 How to construct a results table	AIS2
	33 What to plot? Different types of graph	AIS1

Skill type	Skill	Links to activities
Maths	34 How to plot a line graph	AIS1, AIS2
	35 Reading graphs	AIS1, AIS2
	36 Correlations and scattergrams	A15
	37 Logarithmic scales	A10
	38 Sampling	A14
	39 Hardy-Weinberg equilibrium	A8, A9
	40 Statistical tests and values of p	AIS3
	41 Choosing the right statistical test	AIS3
	42 Normal distribution and standard deviation	A11
	43 Standard error with 95% confidence limits	A11
	44 The chi-squared test	AIS3
	45 Statistics – the t-test	A11
	46 Statistics – the Spearman rank test	A7

Activities

A1 DEVELOPING A NEW ORAL REHYDRATION THERAPY

Diarrhoea is a very common killer in the developing world. Bacterial diseases such as cholera, typhoid and dysentery are all caused by drinking contaminated water. These diseases can be fatal very rapidly because the body becomes dehydrated. As well as water, vital ions such as sodium, chloride and potassium – collectively called electrolytes – reach dangerously low levels. Children and the elderly are at highest risk.

Ideally, treatment involves giving the patient oral rehydration solution (ORS) and antibiotics. The ORS replaces lost water and electrolytes and also provides some sugar for energy. The antibiotics kill the bacteria. It is simple to make an oral rehydration treatment – eight teaspoons of glucose and one of salt dissolved in a litre of water will be very effective. The World Health Organization (WHO) gives the following formulation for an oral powder: "Sodium chloride 2.6 g, potassium chloride 1.5 g, sodium citrate 2.9 g, anhydrous glucose 13.5 g. To be dissolved in sufficient water to produce 1 litre."

△ Fig 26 Cholera is an infection of the small intestine that is transmitted to humans through contaminated food or water.

New ORSs are being formulated all the time. A company wanted to test its new ORS formulation. Following an earthquake in the Caribbean, a team of health workers travelled to the area to test the new formula against the current most effective solution. They carried out a double-blind study of 800 people, measuring the number of patients who needed further attention, such as intravenous treatment.

QUESTIONS

1. Calculate how much glucose would be needed to make up 200 ml of the WHO formulation.
2. The solutes for the WHO ORS would come in a sachet of powder weighing 20.5 g. What percentage of this is glucose?
3. The body needs to replace the lost glucose and salts. Explain why it would not be an advantage to the patient to increase the amount of salt and glucose in the ORS.
4. Many of the 800 people in the study group were children. Suggest an ethical issue involved in using children in clinical trials.
5. **QWC** Describe and explain how the trial should be carried out. Your answer should include an account of how a double-blind trial is done, and how the 800 people were divided into two groups.

Skills practised
3, 6, 7, 8, 9, 10, 18–22, 25, 27

A2 TESTING A NEW FERTILISER

Plants have a few basic needs: light, water, carbon dioxide, a suitable temperature and a few mineral ions. The purpose of fertiliser is to make sure that mineral ions such as nitrate, phosphate and potassium are not limiting factors. Some fertilisers are marketed as plant 'food' but this is misleading. In reality, plants make sugars via photosynthesis and then use the mineral ions absorbed through the roots to make more complex substances such as proteins, nucleic acids and phospholipids.

A pharmaceutical company has developed a new plant fertiliser – Bloomin' Marvellous – for use in domestic gardens. They wanted to advertise it as making plants grow 30% more than the current best-selling brand.

△ Fig 27 In hydroponics, plants can be grown without soil. Instead, the roots are immersed in a solution with just the right amount of nutrients. .

When introducing the new fertiliser, advertising standards rule that any claims must be backed up by evidence, so the researchers at Bloomin' Marvellous had to design and carry out a valid trial to demonstrate the powers of the new fertiliser.

They took five species of common garden plants. For each species, they took three identical batches of seeds.

- Group A received the new fertiliser.
- Group B received the current market leader.
- Group C received no fertiliser at all.

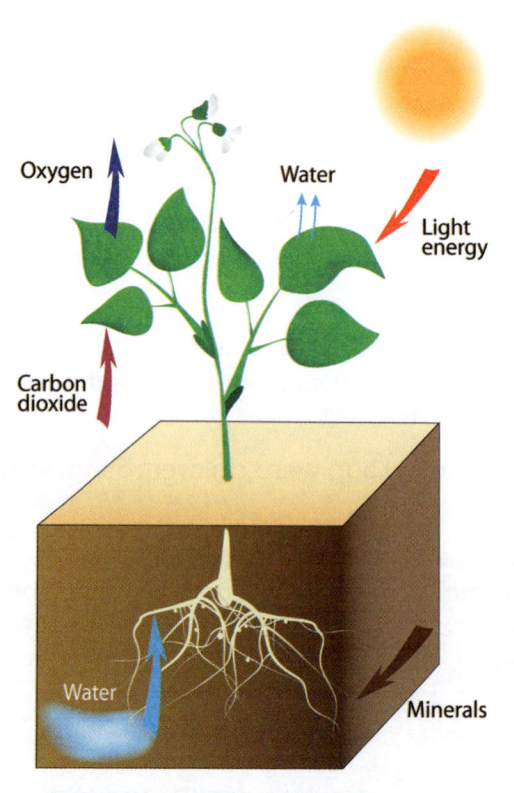

△ Fig 28 An overview of plant function.

All of the plants were grown hydroponically. The roots were suspended in an inert medium instead of soil. After four weeks of identical treatment, the plants were dried in an oven and their dry weight recorded. The results are shown in Table 10.

Average weight of each individual plant (g)			
	Group A	**Group B**	**Group C**
Species 1	55.5	36.8	12.1
Species 2	23.5	20.1	5.5
Species 3	77.3	55.1	10.2
Species 4	34.5	33.3	15.5
Species 5	19.9	17.5	6.6

△ Table 10

QUESTIONS

1. Explain the term "limiting factor".
2. In this investigation:
   a) Name the independent variable.
   b) Name the dependent variable.
   c) Name three variables that should have been controlled in each investigation.
3. Explain why the researchers chose to grow the plants without soil.
4. Which species showed the greatest increase in mass?
5. Suggest why the dry weight of the plants was measured.
6. Suggest the purpose of group C.
7. The plants in group C grew despite having no fertiliser. Suggest a reason for this.
8. **QWC** Evaluate the claim "Bloomin' Marvellous makes plants grow 30% more than the market leader". Use calculations to back up your answer.

> **Skills practised**
> 3, 5, 18–22

A3 HOW CLEAN IS YOUR RIVER?

Contamination of waterways by sewage is a serious health problem. It is illegal to discharge raw sewage into any waterways in the UK but accidents do happen, due to broken sewage pipes, for example. Sometimes animal waste from farmland can wash into waterways. It is important to be able to test for traces of contamination and to assess the extent of the problem.

Bacteria are so tiny that even a water sample that appears clean can contain millions of individuals. Most will be harmless but some will be potentially pathogenic. If you look down a microscope you might see bacteria but there will usually be too many to count. How can you get a meaningful estimate of their true numbers? One way is to perform a **serial dilution**. The basic steps are:

Step 1 – Take 1 cm^3 of the river water and add 9 cm^3 distilled water.

Step 2 – Take 1 cm^3 of this dilution and repeat the process.

Step 3 – Repeat another three times.

Step 4 – Pour the final dilution onto an agar plate and incubate for 48 hours at 25 °C.

- Sample results: after 48 hours there were 16 visible colonies. Each colony had grown from one individual bacterium, so we know that 1 cm^3 of the final dilution contained 16 individuals.

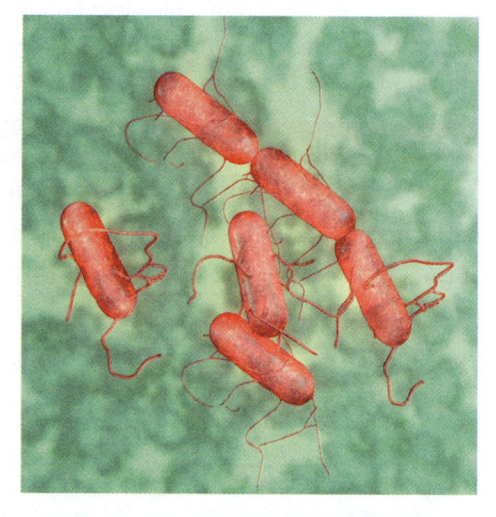

△ Fig 29 Computer generated artwork of bacteria.

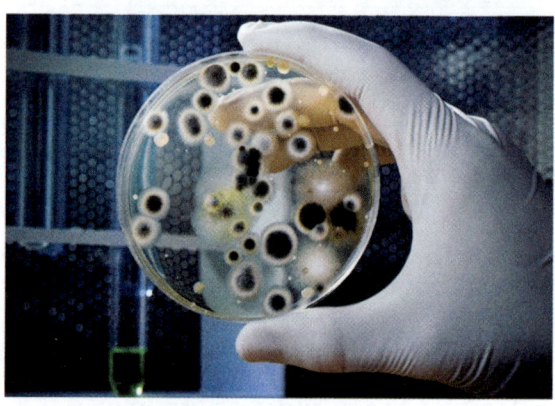

△ Fig 30 Growing bacteria on a Petri dish in the lab.

QUESTIONS

1. After step 1, what was the dilution of the original sample?
2. After step 3, what was the dilution of the original sample?
3. What was the bacterial population, in individuals per cm^3, in the original sample? Explain your answer.
4. When viewed down a microscope, living and dead bacteria look the same. How do we know that this is an accurate estimate of the numbers of living bacteria in the sample?

BOD

Another way to estimate the bacterial population in a sample of water is to use the **biochemical oxygen demand (BOD)** test.

Most bacteria are aerobic – they respire using oxygen. The BOD test can be used to estimate water quality by measuring the amount of oxygen used by the bacteria as they break down the organic matter. It's a good alternative to counting individual bacteria because in most cases it's the lack of oxygen that is directly responsible for a loss of species diversity in contaminated waterways that have become polluted with sewage, or over-fertilised, which is basically the same thing.

The BOD test protocol:

- Take a sample of water and measure the oxygen content.
- Incubate the sample in the dark for five days at 20 °C.
- Measure the oxygen content again.
- The difference is the BOD.

△ Fig 31 Sewage pouring out of a sewer pollutes a lake.

QUESTIONS

Sample	BOD mg O_2 l^{-1}
Pristine, unpolluted river	Below 1
Moderately polluted	2–8
Treated sewage	20 or less
Raw sewage	200–700 or more

△ Table 11 Sample results

5. What is a protocol?

6. Explain the units *mg O_2 l^{-1}* in words.

7. Explain why the sample needs to be incubated in the dark.

8. A student took the following readings, using an oxygen probe.

   Initial reading: 21.5 mg O_2 l^{-1}

   After incubation: 15.4 mg O_2 l^{-1}

   a) Calculate the BOD.
   b) Describe the level of purity of this water sample.

9. There are two components to BOD; both combine to give the total BOD:

   carbonaceous (cBOD) – the proportion of the BOD used in decomposing carbon-based substances such as sugars and cellulose

   nitrogenous (nBOD) – the proportion of the BOD used in decomposing nitrogenous substances such as proteins and ammonium.

   Give a simple equation that shows how to work out cBOD using the values for BOD and nBOD.

10. **QWC** Outline the essential features of the process of eutrophication.

Skills practised
3, 5, 18–22

A4 THE DISCOVERY OF VIRUSES

In science, progress is often limited by the technology available. Research during the late 19th and early 20th centuries that hinted at, then finally identified, the existence of viruses illustrates this point nicely.

Viruses are described as *infectious agents* rather than living things. They can only reproduce inside living cells and don't carry out the other processes we associate with life, such as feeding and respiring. They are so small that even the best light microscopes can't show them. While viruses were discovered in the 19th century, the first images were not available until the invention of the electron microscope in the 1930s.

How can something be discovered before it can be seen? You have to look at the *evidence*.

In 1884 Louis Pasteur and Charles Chamberland tried to find the cause of rabies. They made a very fine porcelain filter that would catch any cells and bacteria but failed to isolate the cause of the disease.

In 1892 the Russian biologist Dmitri Ivanovski tried to find the cause of a plant disease, tobacco mosaic. He took extracts from infected plants and used a Pasteur-Chamberland filter. The filtrate still caused the disease, which led to the hypothesis that it could be due to a toxin, or that there could be infective agents even smaller than bacteria.

In 1898 the Dutch microbiologist Martinus Beirjerinck observed that the infectious agent could multiply inside living cells, disproving the toxin hypothesis. However, it was still thought that the infectious agent was a sort of self-replicating fluid. In the early 20th century, the work of Wendell Stanley showed that viruses were actually tiny particles.

The first images of viruses were obtained from the newly invented electron microscope in 1931 by the German scientists Ernst Ruska and Max Knoll.

So far, over 5000 different types of virus have been discovered. The smallest one is just 20 nm in diameter, while the largest is over 400 nm.

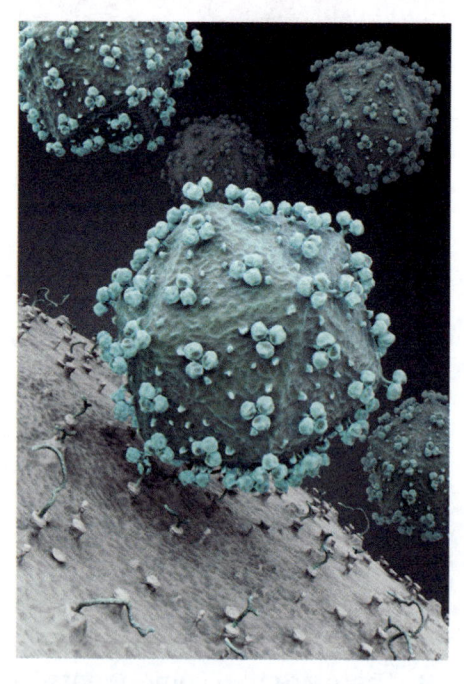

△ Fig 32 3D computer artwork of an HIV particle attaching to a host cell. cell.

QUESTIONS

1. What is a hypothesis?
2. Give two examples of hypotheses that arose from the investigations described.
3. What is a filtrate?
4. Bacteria can be cultured on agar plates but the filtrate could not. Explain why.
5. What was the key piece of technology that showed that rabies and tobacco mosaic are not caused by bacteria?

6. Put the following in order of size, largest first:

   6 m, 66 cm, 226 nm, 846 mm, 2000 µm

7. Put the following measurements into standard form (i.e. give the answer in the form of 5.5×10^x m):

   a) 55 nm
   b) 55 mm
   c) 55 µm

8. Change the following measurements into more suitable units:

   a) 0.0035 mm
   b) 3500 nm
   c) 4655 µm
   d) 0.000065 mm

9. Explain the difference between magnification and resolution.

10. a) Explain why the resolution of electron microscopes is much higher than that of light microscopes.
    b) A particular electron microscope has a resolution of 5 nm. What does this mean?

11. Give simple equations for calculating the following:

    a) actual size
    b) observed size
    c) magnification

12. a) The mitochondrion shown in Fig 33 has been magnified 55 000 times. What is its actual size?
    b) A student used centimetres instead of millimetres when calculating the actual size of an object. How will this affect her answer?

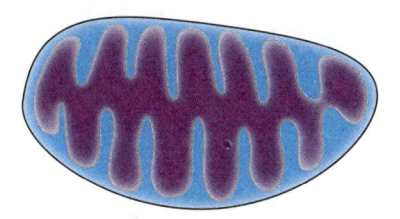

△ Fig 33 Mitochondrion. The image was drawn from a micrograph that had been enlarged 55 000 times.

13. A scientist once asserted, "The microscope is the most important invention in the history of science."

    a) What is an **assertion**?
    b) **QWC** Evaluate this assertion.

Skills practised
1, 2, 18–22, 23, 25

A5 HOW IS DIABETES DIAGNOSED?

The word *homeostasis* means "steady state" and refers to the processes that maintain constant conditions within the body. A whole host of different receptor cells constantly monitor physiological levels such as temperature, carbon dioxide, hormones and many more. If a change is detected, a corrective mechanism is activated that continues until the change is reversed. These mechanisms are negative feedbacks, often described as *detection-corrections*.

The control of blood glucose levels is an important aspect of homeostasis. The levels do not have to be constant, but have to be kept within certain limits, usually between 4 mmol l^{-1} and 6.5 mmol l^{-1}.

If a patient has the classic symptoms of diabetes and his fasting glucose levels are high on more than one occasion, this is usually good enough to confirm a diagnosis of diabetes. For borderline cases, however, the oral glucose tolerance test (OGTT) can be used. Having fasted overnight, the patient has his blood glucose level taken and is then given a standardised glucose drink. The amount given is usually 75 g dissolved in water but in recent times it has been found more convenient to ask the patient to drink 394 ml of original flavour Lucozade™. The blood glucose level is taken again after two hours.

The fasting glucose levels should be less than 6.1 mmol l^{-1} and repeated fasting readings of over 7 are usually a diagnosis of diabetes. Two hours after taking the glucose drink the value should be no higher than 7.8 mmol l^{-1}. A reading of over 11.1 mmol l^{-1} leads to a pretty certain diagnosis of diabetes.

The graph in Fig 34 shows the blood glucose levels of two patients, A and B, taken over a 48-hour period.

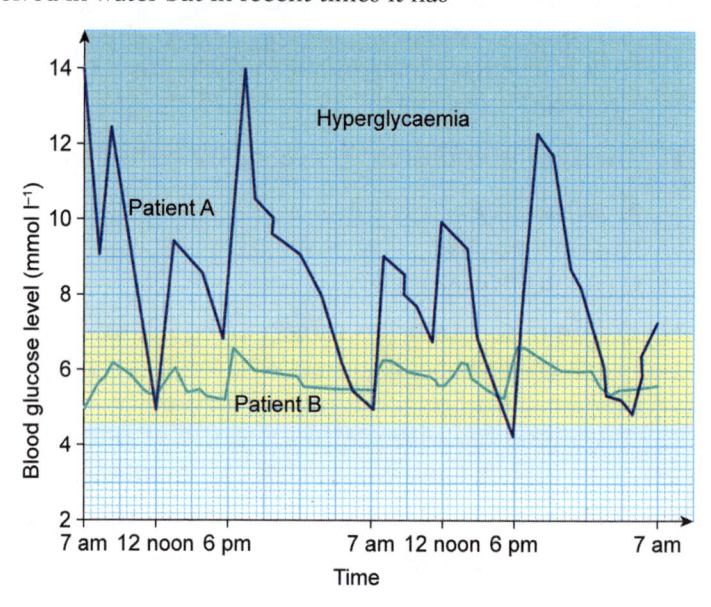

△ Fig 34 Blood glucose levels of two patients taken over a 48-hour period. The yellow region shows the normal range of values.

QUESTIONS

1. The standard international unit for measuring blood glucose is mmol l^{-1}.

   a) *Milli* is a standard prefix in SI units. What is a millimole?
   b) Explain l^{-1} in words.
   c) The relative molecular mass (M$_r$) of glucose is 180 g/mol. A man has his blood glucose measured at 5 mmol l^{-1}. How many grams of glucose would be in one litre of his blood?

2. Explain why patients need to fast before their blood glucose level is taken.

3. Suggest the advantage of asking patients to drink Lucozade™.

4. **QWC** Describe and explain how the body maintains blood glucose levels within limits.

Skills practised
18–22, 24, 25

A6 HOW QUICKLY CAN YOUR BODY COPE WITH SUGAR?

All the cells in the body need a constant supply of glucose for respiration. The problem is that we don't provide our bodies with a constant supply. Sometimes we starve it for hours on end, and sometimes we feel the need to bombard it with three pieces of Black Forest gateau, a whole packet of biscuits and two Mars bars, washed down with two pints of chocolate milk.

In such situations, the body's homeostatic mechanisms are stretched to their limits. Blood glucose levels rise rapidly. The rise is detected by the pancreas, which makes insulin as quickly as possible. Insulin allows glucose to leave the blood and enter cells, stimulating the cells of the liver and muscles to make glycogen at a furious rate. Once the insulin system has done its work, the blood glucose will fall again and, as a consequence, we will feel hungry again, despite the fact that we have more than enough stored energy for our immediate needs.

It is a good idea to have some foods in your diet that release glucose slowly. Beans, for example, contain complex carbohydrates, mainly starch, locked up inside an indigestible case. The advantage of this is a slow absorption that is more closely matched to our needs; less insulin is needed and hunger does not return so quickly. The speed at which glucose is released from any particular food is known as the *glycaemic index*.

Not surprisingly, the food with the highest glycaemic index is glucose, which is given a reference score of 100. Glucose does not need digesting and so is rapidly absorbed as soon as it reaches the region of the gut capable of glucose absorption, the small intestine. All other foods have a GI of less than 100. To work out the GI, a 50 g portion of food is eaten and its effect on blood glucose is measured over three hours. This is compared with the effect of eating 50 g of glucose.

Table 12 shows the glycaemic index of some common foods.

Class of GI	Range of GI values	Examples
Low	55 or less	cherries 22, All-Bran™ 44, banana 55
Medium	56–69	basmati rice 58, raisins 64, wholemeal bread 69
High	Over 60	mashed potato 70, steamed white rice 99

△ Table 12

Glycaemic load is often a more usable value than glycaemic index, because it takes into account the amount of food eaten.

The formula for glycaemic load is: $\dfrac{GI \times g \text{ of carbohydrate}}{100}$

QUESTIONS

1. Explain why GI values have no units.
2. Work out the glycaemic load for
    a) 150g of basmati rice
    b) 120g of steamed white rice.
3. **QWC** Outline the key differences between monosaccharides, disaccharides and polysaccharides.

Skills practised
18–22

A7 PARASITES

Statistics: using the Spearman rank test

In parasitic relationships, one species – the parasite – benefits in some way while the other – the host – suffers some harm to a greater or lesser extent. It is a harsh but simple biological fact that the vast majority of wild animals are infested with parasites. Fleas, tics, intestinal worms – it seems that all large organisms are under attack from smaller ones that want a piece of them. Parasites evolve with their hosts so that in most cases a host can carry a certain population of parasites without showing any major ill effects. So is it true that the larger the organism, the more parasites it can carry?

The Spearman rank test is used to assess the degree of correlation between two ranked sets of data. Ranked data has an *order* to it: biggest to smallest, lightest to heaviest, and so on. So Spearman rank is suitable for use on data resulting from hypotheses like *Students who are good at science will also be good at maths* or *The more fat in the diet, the higher the incidence of heart disease*.

△ Fig 35 Almost all wild animals have parasites, both inside and outside their bodies. The flea is a common parasite and most individuals can support a population of fleas without showing any ill effects.

QUESTIONS

In this example, a researcher investigated the hypothesis *The larger the mouse, the more fleas it will have.*

Step 1 Collect your data

You would gather two sets of data. Data set 1 is the weight of the mouse and data set 2 is the number of fleas on each individual mouse.

1	2	3	4	5	6
Weight of mouse (g)	Number of fleas	Rank of data set 1	Rank of data set 2	d (column 3 minus column 4)	d^2
15.3	12				
18.0	18				
21.5	22	1	1		
20.2	19				
16.7	14				
17.9	15				
19.8	17				
19.0	18				
18.8	20				
17.2	13				

△ Table 13

Step 2 Rank the data

The heaviest mouse is rank number 1 and the greatest number of fleas is rank number 1. As there are 10 measurements, the last rank is 10.

1. Fill in the missing values in columns 3 and 4.

Step 3 Work out values of *d* and *d²*

2. Work out the values in column *d*, where *d* is the difference between the rank in data set 1 and the rank in data set 2.

3. Work out the values of *d²*.

4. Suggest the advantage of using *d²* over *d* when working out the Spearman rank value.

Step 4 Substitute this data into the formula for Spearman rank

$$r_s = 1 - \frac{6\sum d^2}{n(n^2 - 1)}$$

Where

r_s is the Spearman rank, given the name rho (pronounced row, as in "row the boat")

n is the number of data sets you have.

5. Work out the value of r_s.

Interpreting the value of r_s

All values for Spearman rank fall between −1 and +1. A value of +1 means a perfect positive correlation – in this case, the larger mouse always has more fleas – and a value of −1 means a perfect negative correlation. A score of −1 here would mean the smaller the mouse, the more fleas it carries.

As a rough guide, the value can be placed on a scale like this, and the more towards an extreme your values lies, the better the correlation.

−1 _____0_____ +1

However, looking at a value's position on a line is only a rough guide. To be more specific, we can get a value of p by looking our value up on a table of critical values (Table 14).

n	0.10	0.05	0.025	0.01
4	1.000	1.000	–	–
5	0.800	0.900	1.000	1.000
6	0.657	0.829	0.886	0.943
7	0.571	0.714	0.786	0.893
8	0.524	0.643	0.738	0.833
9	0.483	0.600	0.700	0.783
10	0.455	0.564	0.648	0.745
11	0.427	0.536	0.618	0.709
12	0.406	0.503	0.587	0.678

△ Table 14 Critical values for Spearman rank values

QUESTIONS

6. What is *n* in our example? (It might help to highlight the relevant row.)

7. Our Spearman rank value exceeds the critical value in the 0.01 column. What does this mean?

8. **QWC** Outline the differences between a mutualistic and a parasitic relationship. Illustrate your answer with examples.

Skills practised
18–22, 47

A8 PEDIGREE CATS
Using the Hardy-Weinberg principle: part 1

△ Fig 36 Siamese cat.   △ Fig 37 Burmese cat.   △ Fig 38 Tonkinese cat.

The word pedigree means *of known ancestry*. Many people are now using the internet to trace their own pedigree, or family tree. In the animal world, pedigree cats and dogs have been selectively bred for generations according to breed standards. This is controversial because it can lead to an increase in the number of genetic defects.

Variation within a species is caused by different individuals having different alleles. Alleles are alternative forms of genes. New alleles arise from the process of mutation. Some genes have no alleles, because they code for vital products, such as enzymes, and any mutation would seriously harm the organism. The genes that can mutate without proving lethal are responsible for the variation we see between individuals of the same species. Some genes have two alleles and some have more than two.

An example of a gene with three alleles is seen in pedigree Siamese, Burmese and Tonkinese cats.

- The normal allele, C^c, makes the cat's coat blackish and is dominant.
- Siamese cats have a recessive allele, C^s, which makes their coat pale cream and they must be homozygous for this allele.
- Burmese cats have a different allele of the gene, C^b, which makes the coat colour dark brown instead of pale cream. The extremities are black as in Siamese cats. Burmese cats must be homozygous for the C^b allele.
- Tonkinese cats result from the genotype $C^b C^s$; they have a pale brown coat that is intermediate between the Siamese and the Burmese.

QUESTIONS

1. What does the Hardy-Weinberg principle predict about allele frequency from one generation to the next?
2. What does the word pedigree mean?
3. Would you expect the Hardy-Weinberg principle to apply to pedigree cats? Explain your answer.
4. If the frequency of the C^c allele is 0.75, and the C^b allele is 0.1, what is the frequency of the C^s allele? Explain your answer.
5. a) What is a gene pool?
   b) What does pedigree breeding do to the gene pool?
6. **QWC** Explain why the breeding of pedigree cats and dogs towards breed standards is controversial.

Skills practised
18–22, 40

A9 WHO'S HIDING A RECESSIVE ALLELE?
Using the Hardy-Weinberg principle: part 2

By definition, a recessive allele is one that is only expressed in the absence of the dominant version. In any population there will be lots of individuals who are carrying a variety of recessive alleles and there's no way of knowing which because they all look the same. However, if you know how many individuals are homozygous for a particular recessive allele, the Hardy-Weinberg equations allow you to work out how many heterozygotes there are in the population as well.

For example, in a species of guinea pig, the allele for short hair, H, is dominant to allele h for long hair. In a population of 500 animals, 80 were long-haired.

△ Fig 39 Long-haired guinea pig.

Why it's good to be diploid

When an allele mutates, the base sequence changes so that it no longer makes the functional protein that the individual needs. Such alleles are always recessive because they simply don't work. Their effect will be hidden by the allele that does work normally, and which by definition is dominant. The effect of these faulty alleles is only seen when an individual carries two of them. Cystic fibrosis is a classic example.

△ Fig 40 Short-haired guinea pig.

It is a sobering thought that most of us almost certainly carry some mutated alleles. Fortunately, we are diploid – we have two copies of every allele – and the other copy will function normally. If it didn't, you probably wouldn't be here, or would suffer from some genetic condition to a greater or lesser extent. Most of us never get to know which faulty alleles we possess because, if we reproduce, we generally do so with an unrelated individual. This is called outbreeding and it is a wonderful thing. Unrelated individuals will have different faulty alleles and so there is little chance of these being paired up and expressed.

QUESTIONS

1. What is the genotype of the long-haired guinea pigs?
2. How many alleles for hair length are contained in this population?
3. Use the Hardy-Weinberg equation to predict the number of guinea pigs that are heterozygous.
4. In another population of wild guinea pigs, a mutation gave rise to an allele that produced some albino individuals. Predict what will happen to the frequency of this allele. Explain your answer.
5. **QWC** Outline the conditions necessary for the Hardy-Weinberg principle to hold true.

Skills practised
18–22, 40

A10 PESTICIDES, POLLUTANTS AND FOOD CHAINS

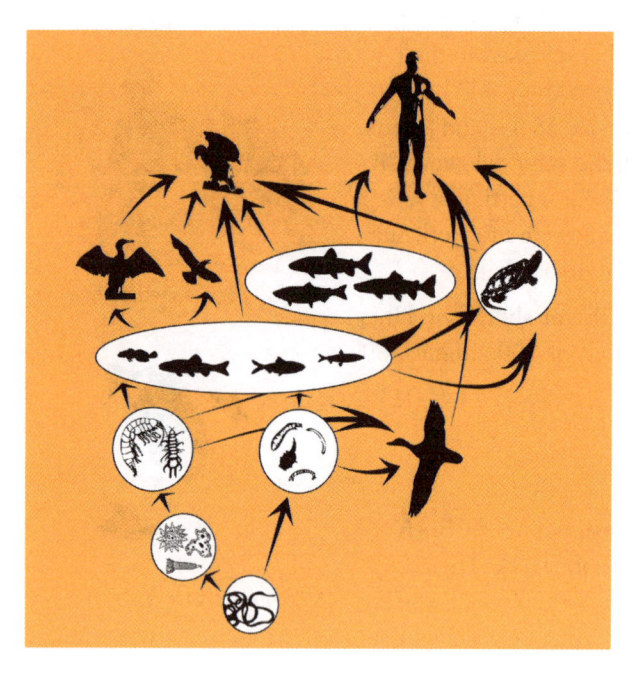

△ Fig 41 Pesticides can accumulate up food chains, with the species at the top being the most affected.

There is a wide variety of toxins that can find their way into ecosystems. Some of them are pollutants, by-products of various industrial processes, and some are pesticides, sprayed into agricultural land to reduce losses. The problem is that some pollutants are fat-soluble, meaning that they dissolve in the fatty tissues of the body. However, they cannot be excreted, because organisms can only excrete water-soluble compounds.

So if a pollutant gets into a lake, for example, it may be present at a very low concentration. Some of it will be absorbed into plants and other photosynthetic organisms, such as algae. When small animals such as water fleas eat the algae, they accumulate all the pollutant from each individual organism they consume. In this way, the pollutant is magnified at each trophic level.

Bioaccumulation refers to the build-up of toxins within an individual organism.

Bioamplification refers to the build-up of toxins as you move up the food chain.

At the top of the food chain, concentrations reach levels that can prove lethal. One particular pollutant, methylmercury, is a by-product of combustion of mercury-containing waste. One study revealed the following levels (see Table 15) in an aquatic ecosystem.

Trophic level	Concentration (ppb)
Fish eating birds (for example herons)	20 000
Predatory fish	500–5000
Small fish – sticklebacks, minnows	200–2000
Zooplankton	50–1000
Algae, aquatic plants and detritus	30–300
In the water	0.02

△ Table 15 Concentration of methylmercury in an aquatic ecosystem.

△ Fig 42 Algal blooms can result from eutrophication – an over-abundance of the mineral ions that algae need. Unfortunately, the same waste that causes eutrophication can also contain high levels of pesticide that accumulate in the algae. Anything feeding on the super-abundant alga will get an extra large dose of pesticide.

QUESTIONS

1. *ppb* means parts per billion. Express 1 billion in standard form.
2. Express the methylmercury concentration in the water in standard form.
3. Higher concentrations are often expressed in terms in the part per million (ppm). Express 20 000 parts per billion as parts per million.
4. By how many times has the methylmercury been amplified from the water to the heron?
5. Methylmercury has a half-life of about 72 days. If a sample of detritus has a level of 200 ppb, calculate what the level will be after 216 days.
6. Computers can be used to predict the effect of introducing pesticides and pollutants into ecosystems. The effect of particular substances on the populations of organisms at the different trophic levels can be estimated.
   a) What is the name given to this process?
   b) Suggest why this process might be inaccurate.
7. **QWC** It has been argued that the extinction of a few species here and there is no big deal. Throughout the history of evolution, species have come and gone, to the extent that the extinct ones far exceed the living ones. The vast majority of these cases have had nothing to do with man. How would you respond to this statement?

Skills practised
17, 18–22, 38

A11 CAN WE CONTROL AN ANIMAL'S GROWTH RATE?

Using the t-test

The problem of how to make animals and plants grow faster has puzzled farmers and scientists for centuries. Selective breeding, also called artificial selection, has brought about remarkable increases in growth rate. By breeding together the fastest-growing individuals for generation after generation, we have produced animals that are ready for market in a fraction of the time it would have taken their ancestors. A new-born piglet weights about the same as a human baby but by the time the piglet is 12 months old it will have reached 90 kg – the weight of a large adult human.

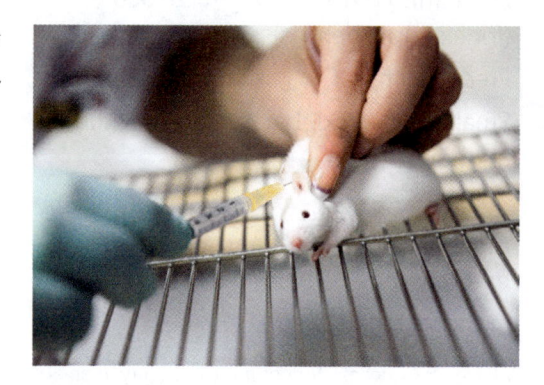

△ Fig 43 Can we control an animal's growth rate?

Scientists have also been researching the use of hormones as a way of increasing growth rate. The thyroid gland secretes thyroxine, a hormone that affects metabolic rate. In an investigation into the effect of thyroxine on the growth of rats, 12 rats were injected with a solution of thyroxine and another 12 acted as a control group.

QUESTIONS

1. Suggest how the control group should have been treated.

2. State a null hypothesis for this investigation.

3. The rats were weighed after four weeks and their change in mass was recorded. If thyroxine was affecting the rats, the control group should have a significantly different mass from the experimental group. Explain what is meant by *significantly different*.

4. Explain how the rats should have been allocated into the control or the experimental group.

5. The results are shown in Table 16.

Experimental group (g)	Control group (g)
3.1 3.0 4.0 4.1 3.8 4.0	3.0 2.7 3.0 2.6 2.7 2.8
3.5 2.9 2.9 3.5 4.2	2.5 2.4 2.4 2.5 2.9 3.0

△ Table 16

The t-test can be used to find out if the two sets of data are significantly different. It involves first calculating the mean and the standard deviation of the data. Explain how the mean should be calculated.

6. The mean was calculated, giving the results in Table 17.

	Experimental group	Control group
Mean growth rate (g)	3.55	2.71

△ Table 17

It is clear that the two means are different but are they statistically significant? The next step is to calculate the *standard deviation*.

Explain the difference between a small and a large standard deviation.

7. The formula for standard deviation, σ (the Greek letter sigma), is

$$\sigma = \sqrt{\frac{\Sigma(x - \bar{x})^2}{n}}$$

σ = standard deviation
Σ = sum of
x = each value in the data set
$\bar{x}$ = mean of all values in the data set
n = number of values in the data set

Step by step, this formula:
- takes the mean of the values in the group
- for each individual value, subtracts the mean and squares the result
- calculates the mean of all the squared values
- takes the square root of that mean (this is the standard deviation).

Computers, calculators and spreadsheets will all calculate standard deviation very quickly. In the example of our rats, the results were as follows (Table 18).

Group	Standard deviation
Experimental	3.55 ± 0.48
Control	2.71 ± 0.22

△ Table 18

Describe what is meant by *a standard deviation of 3.55 ± 0.48*.

8. The t-test also takes into account the standard error of the two sets of data. Standard error (SE) is a measure of the variability of the data and is worked out by the equation

$$SE = \frac{\sigma}{\sqrt{n}}$$

where σ = the standard deviation of the sample and n = the number of observations in the sample (11 for the experimental group and 12 for the control group).

Calculate the standard error of both groups.

The standard error can be used to generate **confidence limits** on a simple bar graph (see Fig 44). The error bars are worked out as being twice the standard error above and below the mean (actually, it's 1.96 times the standard error but it can almost always be rounded up to 2). The golden rule is that if the error bars do not overlap, we can be more than 95% confident that the results are not due to chance. The results are significantly different. But what if the error bars do overlap a little? Does that mean the results are not significant? The simple answer is that they might still be significantly different or they might not and this is where the t-test can help.

The t-test will calculate how big the difference between the two sample means is, relative to the standard errors. This means that we have to calculate the difference between the two standard errors, using the equation

$$SE_D = \sqrt{(SE_A)^2 + (SE_B)^2}$$

The standard error of the difference (SE_D) is worked out by squaring the standard errors of both groups, adding them together and taking their square root. This works out to be 0.152.

Now we're almost there. The formula for the t-test is

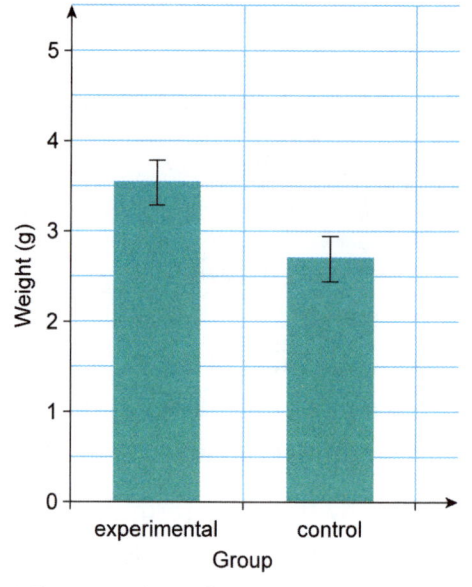

$$t = \frac{\text{mean difference in samples (i.e. experimental – control)}}{\text{standard error of difference}}$$

In terms of the values we already have,

$$t = \frac{3.55 - 2.71}{0.152}$$

$$t = \frac{0.84}{0.152}$$

So t = 5.53

△ Fig 44 Experimental group vs. control.

We now need to decide if this value of t is significant. We look up 5.53 in a table of critical values but we need to know the number of degrees of freedom. This is worked out as the sum of the number of measurements in each category, minus one for each category. So in this case we've got 12 – 1 plus 12 – 1, which is 22.

For 22 degrees of freedom, the critical value of t is 2.08. Our value is easily higher so we can say that the results for our two groups are significantly different. When done on a computer, the results gave a p value of less than 0.0001, which is described as extremely significant. We can be 99.99% certain that our results are not due to chance

9. The investigator wrote up his findings and sent them for publication in a journal, under the title "Thyroxine shown to increase growth rate in rodents".

   a) What is the purpose of journals?
   b) What is the role of **peer review**?
   c) **QWC** Evaluate the investigator's statement, "Thyroxine shown to increase growth rate in rodents."

> **Skills practised**
> 4, 5, 11, 12, 15, 18–22, 43, 44, 46

A12 WHO'S AT RISK OF HEART DISEASE?

The underlying cause of coronary heart disease is a narrowing of the coronary arteries that supply the heart muscle with vital oxygen and nutrients. The narrowing is due to a fatty material called atheroma, which accumulates under the endothelium of the arteries, narrowing the space available for blood flow and making the walls rougher. A crisis occurs when the rough walls cause blood clots to develop that can block blood flow to an area of cardiac muscle.

Coronary heart disease is the single biggest killer in many developed countries. It is not a communicable disease – you can't catch it. The underlying cause is a combination of lifestyle and genetic factors that took us years to piece together. Now that we have a very good understanding of its causes, we are in a position to predict where and when the disease will strike and take steps to prevent it.

Epidemiology is the branch of medical science that is concerned with the incidence and causes of disease. It has an absolutely vital role to play in public health because it identifies risk factors and allows effective strategies for preventative medicine.

Epidemiological studies have been conducted on all the major health risks in western countries. The aim is to be able to predict the likelihood of somebody developing a particular condition, such as coronary heart disease, diabetes or cancer. The best way to do this is a cohort study.

A cohort is a group of people who have something in common, such as year of birth, or exposure to a particular environmental factor, such as a toxin. A cohort study takes a group of people and gathers data about them over a prolonged period, sometimes over generations. The idea is to gather information about which individuals develop the disease and match this information to a study of their lifestyle.

A cohort study into the factors affecting the development of coronary heart disease resulted in the following scheme (Table 19), aimed at creating a points score that will predict an individual's risk of having a heart attack in the next decade.

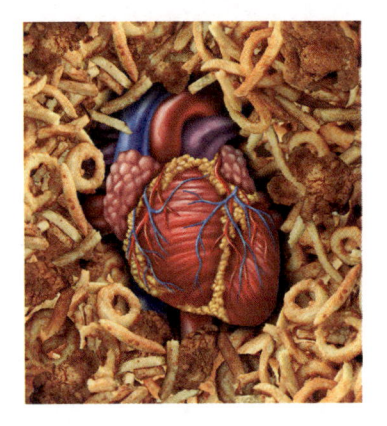

△ Fig 45 Foods rich in saturated fat and cholesterol contribute to the build up of atheroma in all the arteries. Atheroma in the coronary arteries restricts blood flow to the heart muscle.

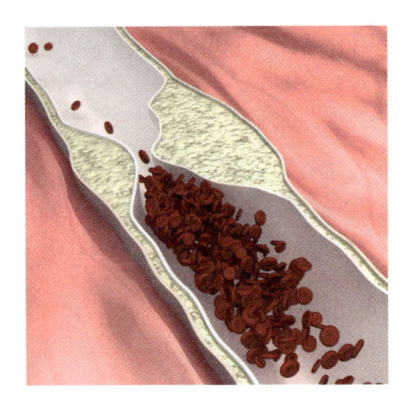

△ Fig 46 The build up of atheroma, shown here in white, restricts the flow of blood in arteries and also roughens the lining of the vessel, which makes blood clots more likely. Here, a blood clot has completely blocked the vessel.

| Factor | Range | Points for each age group ||||
| | | Age group (decade of life) ||||
		Forties	Fifties	Sixties	Seventies
Age		3	7	10	12
Blood cholesterol mg 100 cm⁻³	<160	0	0	0	0
	160–199	3	2	1	0
	200–239	5	3	1	0
	240–279	6	4	2	1
	280+	8	5	3	1
Smoking	Smoker	5	3	1	1
	Non-smoker	0	0	0	0
Blood pressure (systolic) mm Hg	<120	0	0	0	0
	120–129	0	0	0	0
	130–139	1	1	1	1
	140–159	1	1	1	1
	160+	2	2	2	2
Blood HDL content mg 100 cm⁻³	<40	2	2	2	2
	40–49	1	1	1	1
	50–59	0	0	0	0
	60+	−1	−1	−1	−1

△ Table 19 Scheme to predict the risk of heart attack.

Points score	1	2	3	4	5	6	7	8	9	10	11	12	13	14	15	16	17+
% increased risk of CHD	1	1	1	1	2	2	3	4	5	6	8	10	12	16	20	25	30

△ Table 20 Points score for men and the percentage probability that they will develop CHD in the next 10 years.

QUESTIONS

1. What does the sign < mean (as in <120)?

2. Some people are *genetically predisposed* to heart disease.
   a) What does genetically predisposed mean?
   b) Suggest why genetic predisposition does not feature as a risk factor in Table 19.

3. Identify three trends from Table 19.

4. Each year there are around 600 000 deaths in the UK, of which around 26% are due to coronary heart disease. Calculate how many people die each year from diseases of the heart and circulation.

5. a) Work out the 10-year risk for a smoker, age 51, who has a blood cholesterol of 255, a systolic blood pressure of 145 and an HDL level of 55.
   b) **QWC** If you were this patient's GP, what advice would you give him about changing his lifestyle?

Skills practised
18–22

A13 THE POTENTIAL OF STEM CELLS

A human begins life as one single fertilised cell – a zygote – and grows into a massively complex body that is built from over 200 different types of specialised cell. The early cells of the embryo are stem cells, which have two vital properties:

- self-renewal – they can divide again and again without becoming differentiated
- potency – the ability to differentiate into specialised cells.

There are three main types of stem cell, classified according to their abilities.

△ Fig 47 Artwork of an embryo at the 32 cell stage.

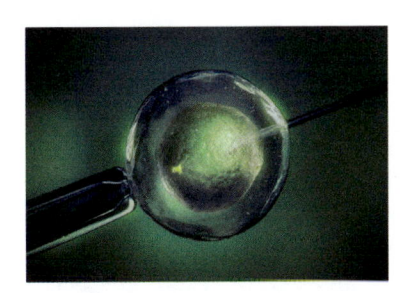

- Totipotent stem cells. These can give rise to whole organisms. Only the very early embryo contains these cells.
- Pluripotent stem cells. These come from the *inner cell mass* of the embryo and have the potential to differentiate into any of the cell types that make up the human body. The difference between these and totipotent cells is that cells from the inner cell mass cannot generate a whole new organism, because they can't differentiate into extra-embryonic tissue (such as the placenta), which is needed to support a foetus.
- Multipotent stem cells are more limited and can differentiate into just a few closely related types of cell. Hematopoietic stem cells in bone marrow, for example, can differentiate into red blood cells, different types of white blood cell and platelets. Most adult stem cells are of this type.

Most scientists agree that stem cells offer great hope for the future of regenerative medicine. This was seen as problematic in the recent past, because the first two types of cell are only found in embryos, which are destroyed in the process of harvesting them. There was a ready supply of human embryos because the process of in vitro fertilisation (IVF) generates more than are needed, but many people were unhappy about their use.

However, nowadays much of the work on stem cell/regenerative therapies has moved away from using embryonic stem cells to induced pluripotent stem cells generated from adult cells. There are banks of "approved" embryonic stem cells that researchers can access, reducing the need to keep using new embryos. The wonderful thing about stem cells is that they replicate indefinitely, so once you have a stock of them in repositories you don't have to keep destroying embryos.

△ Fig 48 In PGD, a cell from an embryo can be taken and its genetic constitution analysed. This does the embryo no harm and if it turns out to be healthy, it can be implanted. .

In cases where parents know they are at risk of having a child with a serious genetic disorder, IVF gives the possibility of being able to choose a healthy embryo. Following the IVF process, each embryo can be screened to see if it carries the disease. This process is known as PGD: pre implantation genetic diagnosis. Few people have a problem with this process, but some worry that it may be a slippery slope to screening for all sorts of non lethal traits such as intelligence, gender and physique that could lead to "designer babies".

Bioethics is largely concerned with moral problems brought about by advances in biology and medicine. Technology limits what we *can* do but ethics is concerned with what we *should* do.

Bioethics is a vast area that covers everything from abortion, through cloning, stem cells, assisted suicide, to transplantation and human organ trafficking. Ethical objections take many forms, including the following.

- It's a slippery slope. Once we start, we won't be able to stop, so where will it lead?
- We don't know enough to predict the long-term effects of interfering with an organism's genome.
- Do the potential benefits outweigh the suffering?
- Human life is sacred. Even a single-celled embryo is a human life and should therefore be afforded rights.

QUESTIONS

1. Explain what is meant by the term *stem cell*.
2. When talking about stem cells, what is meant by the term *potency*?
3. Explain how stem cells can differ in their potency.
4. List three conditions that could potentially be treated successfully with stem cells.
5. **QWC** "Embryonic stem cell research should continue." Evaluate this statement.

Skills practised
6, 18–22

A14 WHERE HAVE ALL THE UK'S FORESTS GONE?

If left to its own devices, a patch of bare ground will develop into a deciduous forest in a few decades. Once it has been left for a hundred years or so, the balance of dominant tree species will not change. At one time, most of the lowland areas of the UK were covered in deciduous forest. Now, there is very little left and virtually none in England.

Forests develop by a process of colonisation and succession. However, this will not happen where there is a significant amount of grazing. Rabbits, sheep, cows, goats and lawnmowers all cut off the growing points of young plants. Many species, including all tree and shrub species, cannot cope and simply die. Some plants have evolved to cope with grazing, however. Grass has its growing point at the base of the stem, so if the tip is chopped off it keeps on growing. The other strategy that can withstand mowing or grazing is to be very small or very flat. Species such as clover and plantain are common on mown grass because of this.

△ Fig 49 When grazing stops, a greater variety of plants will thrive.

When grazing stops, a greater variety of plants will thrive, followed by a greater diversity of insects, birds and mammals. Eventually, a forest develops. The exact nature of the forest depends on the climate of the country.

Lichens are usually the first colonisers of bare rock. These fascinating organisms are in fact not one species but two: an algae growing inside a fungus. This mutualistic arrangement allows lichens to thrive where there is very little water or minerals. The algae photosynthesise while the fungus can prevent drying out, and can provide some essential nutrients. Some fungi even secrete an acid that eats into the rock, releasing a few precious minerals.

△ Fig 50 Life can gain a foothold even in the most inhospitable of places.

Once the lichens have become established, the conditions become more favourable for other organisms such as mosses to out-compete them. In this way, succession can lead to a much more complex ecosystem becoming established.

QUESTIONS

A student carried out an investigation into the effect of mowing on species diversity. She used two fields: one was mowed regularly and the other was left un-mown.

1. For this investigation:
   a) suggest a hypothesis
   b) suggest a null hypothesis.

2. Quadrats are small frames, often 0.5 m square, which allow us to sample patches of vegetation. Describe what is meant by the process of sampling.

3. In order to do a statistical analysis of the results, the quadrats must be placed at random.
   a) What is meant by "at random"?
   b) Describe a method that would allow the quadrats to be placed at random.

4. The student wanted to work out the species diversity on the two patches of land. What two factors combine to give species diversity?

5. Does this investigation need a control? Explain.

Before the investigation, the student needed to know how many quadrats to throw. She threw a number of quadrats on the un-mown area and compiled a list of the species she found. Her results are shown in Fig 51.

6. Use the graph in Fig 51 to explain why:
   a) six quadrats would not be enough for this investigation
   b) 25 quadrats would be too many.

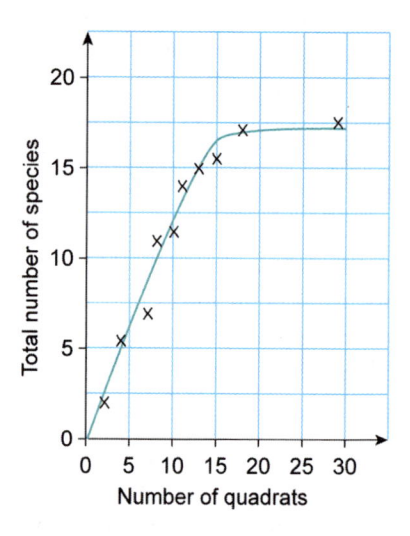

△ Fig 51

7. Suggest how many quadrats the student should include in her study.

8. **QWC** Energy is transferred through an ecosystem. Describe how and explain why the efficiency of energy transfer is different at different stages in the transfer.

Skills practised
18–22, 39

A15 OVERWEIGHT OR JUST BIG-BONED?

Body fat is a difficult topic. What does body fat mean and how much is too much?

All living tissue is made from cells, which are basically systems of membranes, all of which are made from lipids. It's impossible to have no body fat and even low amounts of fat are not healthy. However, health problems begin when we have too much body fat.

Excess fat is stored in adipose tissue, consisting of cells that can absorb triglycerides and simply swell and swell. There are several ways to measure body fat:

- The simplest is probably to look at pictures of people of known body fat composition and use them to estimate our own value. This is cheap and remarkably accurate.
- Pinch an inch. A simple pair of callipers will allow you to measure the subcutaneous (under the skin) fat and read off your body fat percentage from a table.
- More sophisticated – and expensive – techniques include *biometrical impedance analysis* (in which the electrical resistance of your body is taken); measuring the body's displacement in water or air, then using your weight to estimate your density and therefore percentage of body fat; and DEXA scanning, which uses a whole body X-ray scan.

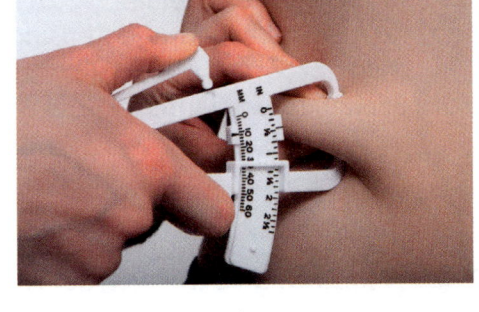

△ Fig 52 Pinching an inch.

QUESTIONS

As part of an investigation into body fat, a class of seven-year-olds were weighed. The results for the boys (in kg) were as follows:

24.4 24.7 22.0 26.4 18.1 33.0 19.2 25.4
24.7 22.3 24.7 23.6 17.1 24.6

1. Rank these weights in order.
2. For these values, give
    a) the mean
    b) the median
    c) the mode.
3. To assess each child's progress, the chart in Fig 53 was used. Use it to explain what is meant by:
    a) percentile
    b) decile
    c) interquartile range.

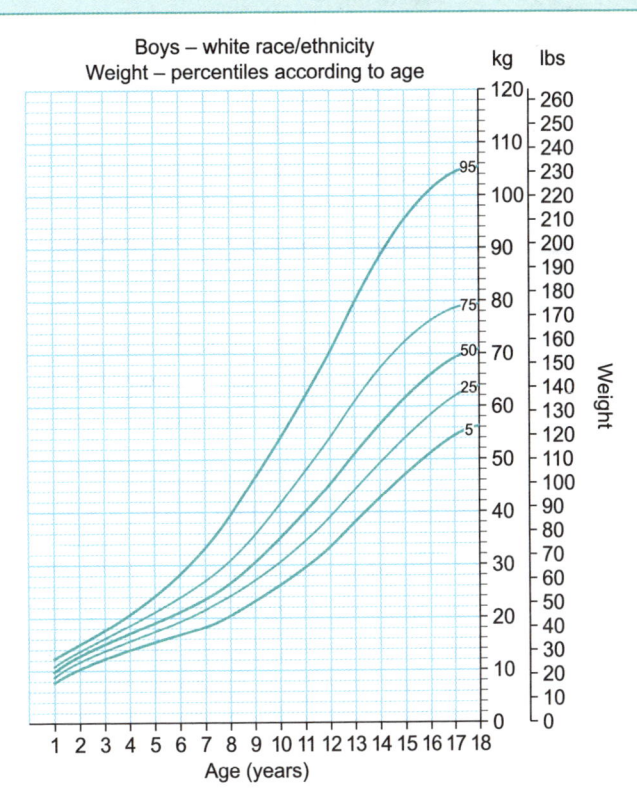

△ Fig 53

4. The scales used to measure the boys were found to be faulty and were reading 5% lower than they should.
   a) What is the name given to this sort of error?
   b) Explain how this sort of error can be eliminated.
   c) Would the data from the faulty scales be classed as imprecise or inaccurate? Explain.
5. a) Explain the advantage of using a chart like Fig 53 to assess a child's progress.
   b) Explain how scientists know where to draw the green lines.
6. Use Fig 53 to assess the comparative weight of:
   a) the heaviest
   b) the lightest child.
7. Use Fig 53 to give the interquartile range of weights.

BMI

A person's weight can be assessed by the body mass index (BMI).

The formula for BMI is

$$\frac{\text{mass in kg}}{(\text{height in m})^2}$$

The classification of the results is usually shown as in Table 21, although the boundaries can vary. For example, some schemes only class a BMI below 18.5 as being underweight.

BMI range	Class
Below 20	Underweight
20–25	Normal
25–30	Overweight
30–40	Obese
Over 40	Morbidly obese

△ Table 21 Typical BMI classification.

QUESTIONS

8. a) Work out the BMI for an individual of 1.84 m tall weighing 90 kg.
   b) How would you classify this individual?

There are problems with the BMI.

Many perfectly healthy teenagers will be classed as underweight. Generally, children have their own BMI charts with different boundaries.

BMI is not the same as body fat. There are many other tissues in the body that contribute to an individual's weight.

In response to the second criticism, scientists measured the body-fat levels of individuals and plotted a graph of BMI against percentage of body fat, shown in Fig 54.

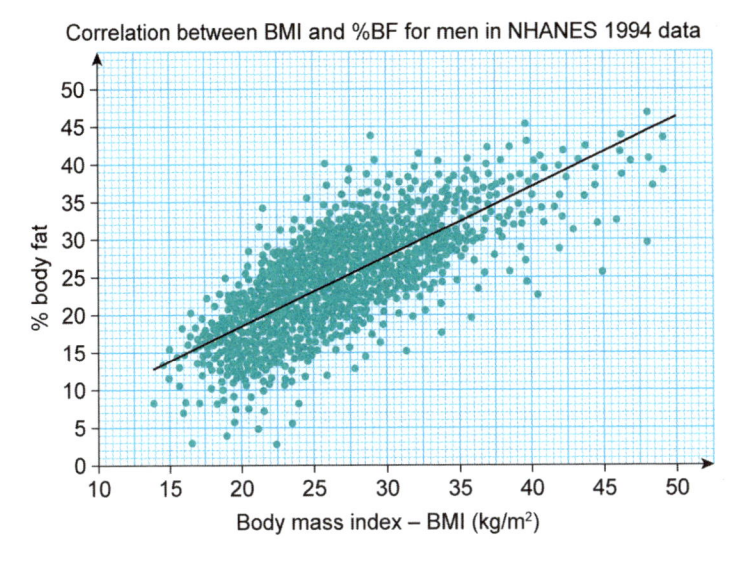

Correlation between BMI and %BF for men in NHANES 1994 data

△ Fig 54

9. What is the name given to the type of graph in Fig 54?
10. The computer drew the line of best fit. What does this line show?
11. a) Circle any individual who has a high BMI but a low percentage of body fat.
    b) Suggest a reason for this apparent **anomaly**.
12. **QWC** In a health study, one health authority wanted to gather information about the lifestyles of everyone in their area. They hoped to compile information on age, gender, BMI, blood pressure and lifestyle in terms of smoking, diet, exercise and alcohol intake. Evaluate how successful and informative such a data-gathering exercise would be.

Skills practised
13, 14, 18–22, 29, 30, 37

Assessing Investigative Skills

AIS1

Investigating the effect of temperature on the activity of trypsin

Trypsin is a protein-digesting enzyme made by the pancreas. It forms part of pancreatic juice, which is secreted on to food in the duodenum. Trypsin's function is to hydrolyse protein molecules into shorter chains of polypeptides.

To investigate how trypsin activity can be affected by temperature – or any other variable – we need to give it some protein to digest. Egg white is almost pure protein. Small glass tubes, open at both ends, can be filled with raw egg white and then boiled so that the protein solidifies. When dipped into enzyme solution the enzymes will digest the egg white from each end. The faster the enzyme activity, the faster the egg white will disappear.

A student chose a range of temperatures to investigate and wrote down the results (Table 22).

Temp (°C)	Time taken to digest protein (min)	Rate (1/time taken) × 100	Rate × 100
10	83	0.012	1.2
20	41	0.024	2.4
30	22	0.045	4.5
40	12	0.083	8.3
50	35	0.028	2.8
60	(protein not digested)	-	-

△ Table 22

In this investigation:

1. What was the independent variable? (1)

2. What was the dependent variable? (1)

3. List three variables that would need to be controlled. (3)

4. Suggest a suitable control. Explain your choice. (2)

5. Explain why this was a suitable range of temperatures. (2)

6. Explain why a line graph was suitable to display these results. (1)

7. A simple way to work out the rate or reaction is to do the reciprocal of time, that is, 1/time. Suggest why *1/time* is better than *time* as a measurement of enzyme activity. (1)

8. Suggest a simple practical reason why rate was multiplied by 100 in the final column. (1)

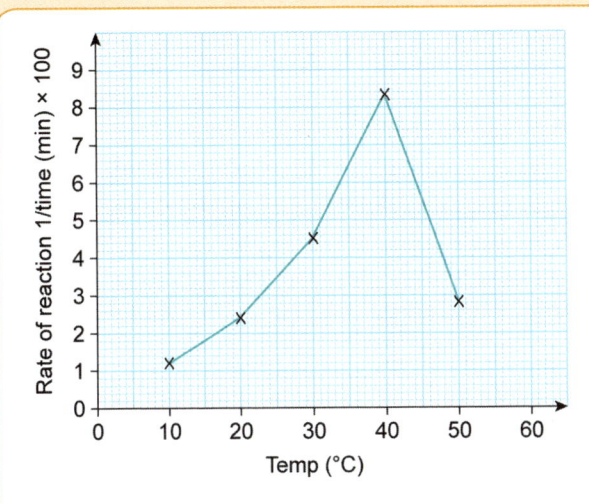

△ Fig 55

9. We can use the graph to estimate the rate of reaction at temperatures for which we took no readings.

   a) What is this process called? (1)

   b) Estimate the rate of reaction at 25 °C. (1)

10. We can also estimate readings by following the line of the graph beyond the range of the readings.

    a) What is this process called? (1)

    b) Predict the rate of reaction at 55 °C. (1)

11. The student concluded that the optimum rate of reaction is at 40 °C. Is this a valid conclusion? Explain. (1)

    The temperature coefficient, known as Q_{10}, is a measurement of the effect of a 10 °C rise on the rate of enzyme activity.

    The formula for Q is: $\dfrac{\text{rate of reaction at } t + 10\ °C}{\text{Rate of reaction at } t\ °C}$

12. Work out the Q_{10} when t is 20 °C. (2)

13. Relate the properties of enzymes to the fact that they are proteins. (7)

Skills practised
3, 5, 18–22, 34, 35, 36

AIS2

Osmosis investigations

Most osmosis investigations involve plant tissue, which will increase in size and weight when placed in a solution of higher water potential and decrease in a solution of lower water potential. This investigation involves placing pieces of potato into a range of sucrose and salt solutions. This will allow us to estimate the water potential of the cytoplasm in the potato cells, see if there is any difference in the osmotic activity of salt and sugar solutions, and relate the concept of water potential to specific concentrations of salt or sugar.

1. Define osmosis in terms of water potential. (2)

2. When sodium chloride dissolves, the sodium and chloride ions separate and both become surrounded by shells of water molecules. This does not happen with sucrose. In which solution would you expect osmosis to occur at the greatest rate? Explain your answer.

The method

Step 1

A student made a range of solutions. He was given a 0.5 M stock solution that needed to be diluted by adding distilled water.

3. Fill in Table 23 with the suitable volumes of water and stock solution. (2)

Concentration (M)	0.0	0.1	0.2	0.3	0.4	0.5
ml salt or sugar	0					20
ml water	20					0

△ Table 23

Step 2

The student used a cork borer to make six identical potato cylinders. He patted them dry with a paper towel, then weighed them individually. He recorded the weights in the results table.

Step 3

Then he put one cylinder into each tube and left them for 60 minutes.

Step 4

He then removed the potato cylinders, dried them and weighed them on the same weighing machine and recorded the results.

4. In this investigation:

   a) What was the independent variable? (1)

   b) What was the dependent variable? (1)

   c) List three variables that were controlled. (3)

   d) Suggest any two factors that could have given inaccurate results. (2)

5. Explain why leaving the cylinders in for a full day would not produce better results. (2)

6. Explain why it is important to use the same balance for weighing, both before and after immersion. (2)

7. The student's results are shown in Tables 24 and 25. Fill in the last two columns in each table. (4)

Concentration (moles)	Initial weight (g)	Final weight (g)	Difference (g)	% Change in weight $\frac{\text{Change in mass}}{\text{Original mass}} \times 100$
0.0	9.58	10.27		
0.1	9.72	10.03		
0.2	9.35	9.29		
0.3	9.90	9.32		
0.4	9.62	8.80		
0.5	9.81	8.57		

△ Table 24 Sample results: salt.

Concentration (moles)	Initial weight (g)	Final weight (g)	Difference (g)	% Change in weight $\frac{\text{Change in mass}}{\text{Original mass}} \times 100$
0.0	10.13	11.04		
0.1	8.60	9.09		
0.2	8.89	9.19		
0.3	9.77	9.85		
0.4	9.27	9.09		
0.5	9.19	8.75		

△ Table 25 Sample results: sucrose.

8. a) Which concentration has the highest water potential? (1)

   b) Which concentration has the lowest water potential? (1)

9. Explain why we need to work out the percentage change, rather than just the difference in weight. (2)

10. Suggest two ways in which the rate of osmosis could be speeded up so that the final weight of the potato was reached sooner. (2)

11. As an alternative to working out percentage change, the results for each potato cylinder can be given as a ratio, worked out as $\frac{\text{initial weight}}{\text{final weight}}$.

    What would the following ratios show about the movement of water?

    a) A ratio of 1.

    b) A ratio of more than 1.

    c) A ratio of less than 1. (3)

    Another student plotted similar results in a line graph.

12. What does the term *isotonic* mean? (1)

13. We can use the graph to estimate the concentration, or molarity, of the cytoplasm of the potato cells.

    a) A ____ M solution of salt is isotonic to potato cytoplasm.

    b) A ____ M solution of sugar is isotonic to potato cytoplasm. (2)

14. Take your answers from question 13 and use the graph below to estimate the water potential of potato cells. (2)

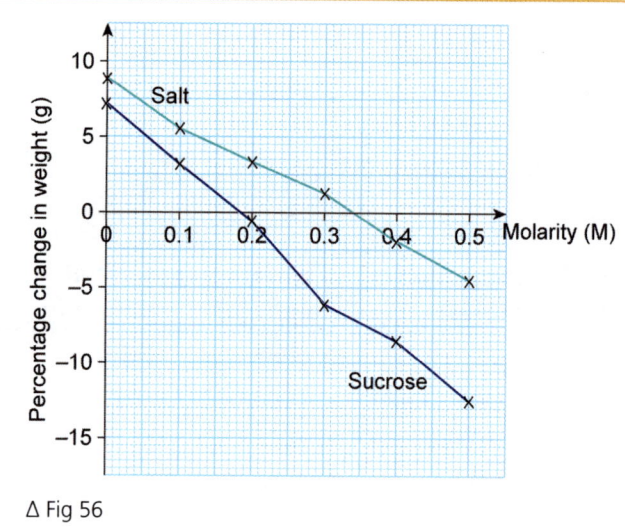

△ Fig 56

The graph allows you to convert molarity (strength of solution) into a water potential value. The y axis is water potential in KPa.

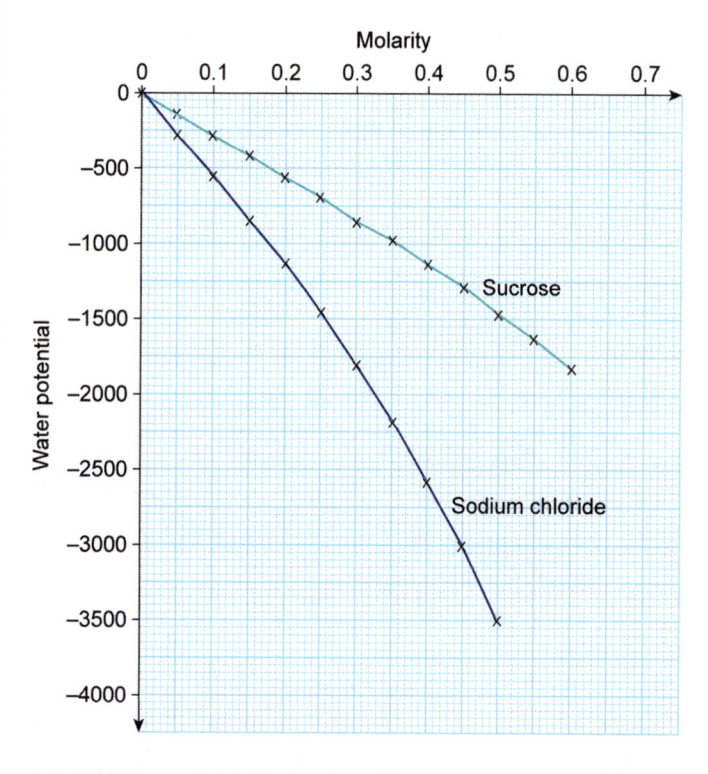

△ Fig 57 Water potential conversion table.

15. Write an essay about the properties of water that are relevant to biology. (25)

Skills practised
3, 18–22, 28, 31, 32, 33, 35, 36

AIS3

Using the chi-squared test

In a coniferous forest a researcher wanted to find out which species of tree, if any, were preferred by red squirrels to build their drey (nest). She gathered the following results (Table 26).

Species of tree	Number of squirrel nests (observed values)	Number you would expect by chance (expected values)
Larch	12	
Spruce	18	
Scots pine	15	
Fir	25	
Total dreys	70	

△ Table 26

△ Fig 58 Red squirrel.

Step 1 Gather your observed values and work out your expected values

The chi-squared test compares the actual data collected – the *observed* values – with those you would expect if there were no significant difference between the categories. This means that the expected value in each category is the mean of all the observed values.

1. Suggest an experimental hypothesis for this investigation. (1)
2. Suggest a null hypothesis for this investigation. (1)
3. Use the observed results to work out the expected values. (1)

Step 2 Work out the value of χ^2

Once we have the observed and expected values, we need to use this equation to calculate the value of χ^2.

$$\chi^2 = \Sigma \frac{(O - E)^2}{E}$$

Where O = observed value

E = expected value

Σ = sum of all

The formula looks complicated but you can break it down into easy stages. If you put the values in a table such as Table 27, each row becomes the next step in working out χ^2.

Species of tree	Larch	Spruce	Scots pine	Fir
Observed results (O)	12	18	15	25
Expected results (E)				
O – E				
(O – E)²				
$\frac{(O-E)^2}{E}$				

△ Table 27 An easy way to calculate χ^2.

4. Fill in all the missing values in Table 27. (5)

We obtain the value of χ^2 by adding together all the values of $\frac{(O-E)^2}{E}$.

5. Work out the value of χ^2. (1)

Step 3 Find out the degrees of freedom

Table 28 shows values of χ^2, but before we can interpret a particular value, we need to know the number of *degrees of freedom*, which is one fewer than the number of categories in the investigation.

6. How many degrees of freedom does this investigation have? (1)

Step 4 Interpret the values of χ^2

We now have a χ^2 value and therefore know the number of degrees of freedom. To find out if our results are significant, we need to know whether our χ^2 value is above or below the critical 5% value. So we need to use Table 28.

7. The critical value of p is taken to be 0.05, or 5%. What does this mean? (1)

Degrees of freedom	Value of p (probability)					
	0.50	0.25	0.10	0.05	0.02	0.01
1	0.45	1.32	2.71	3.84	5.41	6.64
2	1.39	2.77	4.61	5.99	7.82	9.21
3	2.37	4.11	6.25	7.82	9.84	11.34
4	3.36	5.39	7.78	9.49	11.67	13.28

△ Table 28 Chi-squared values

8. You know how many degrees of freedom there are, so highlight the appropriate row. (1)

9. a) Between which two values does our chi-squared value fall? (1)

   b) Between which two values of p does this fall? (1)

10. What is the probability that our results are due to chance? (1)

11. And what is our conclusion about the whole investigation? (2)

Skills practised

3, 4, 41, 42, 45

Answers

Each activity's set of questions gives you the opportunity to practise and develop skills covered in the **Skills** section. During your biology course, you will be asked to analyse, calculate, evaluate and apply your knowledge to new situations. By answering the questions, you will enhance your understanding of the skills necessary for becoming an independent thinker – something examiners are looking for. This section includes all the answers to the activities, as well as helpful hints and tips to boost your performance in the exam room.

The last question in each activity is a quality of written communication (QWC) question. These test your subject understanding and ask you to think about how you communicate your ideas. For these questions, we outline the subject knowledge points you need to cover and also include an indication of low, medium and high level responses to show you how to improve your answers. For more guidance on QWC questions, we have written low, medium and high level responses to Activity 14, Question 8 in the **QWC Worked Examples** section. Each response has detailed points commenting on its quality of written communication, giving you an insight into what's needed.

For more hints on how to improve the quality of your written communication, see the QWC Worked Examples.

The **Assessing Investigative Skills** section gives you the chance to develop your investigative skills in three major activities without the need to complete practical work. You will be able to tackle AIS1 and AIS2 soon after you've started your course and AIS3 will help you tackle a more advanced investigation later on. The answers to these AIS questions offer comprehensive guidance so you can check your responses and see how to progress.

Finally, remember that learning facts is only a small part of studying science. Examiners and future employers are looking for someone who can analyse, think logically and apply their knowledge to new situations. Practising the skills will develop your ability to do this.

A1 Developing a new oral rehydration therapy

1. 2.7 g of glucose (13.5/5).
2. 13.5 as a percentage of 20.5 (13.5/2.6 × 100) = 65.85%.
3. Having more salt/sugar will lower the water potential, so absorption of water into the blood by osmosis will be slower.
4. People involved in trials should ideally give informed consent. Many children will be too young, or too ill, to do so.

5. **QWC** These are the points you should make in your answer:
   - Neither the patients nor the health workers know who is getting the new ORS and who is getting the older one; this eliminates the placebo effect.
   - The two groups should be *randomised*.
   - The two groups should be matched as closely as possible.
   - (Any two examples of matching): age; gender; time since onset of diarrhoea.
   - Give both groups equal volumes of ORS.
   - Give both groups doses at similar intervals.
   - Compare groups in terms of how long symptoms last.
   - Compare groups on terms of how many individuals need further treatment.

   A low level answer would include up to three of these points.

   A medium level answer would include five points.

   A high level answer would include seven or eight points.

A2 Testing a new fertiliser

1. *Either*: Any factor that, if supply is increased, will increase the rate of growth. *Or*: Any factor that, if not available in sufficient amounts, will result in decreased growth rate.
2. a) The type of fertiliser
   b) The growth of the plant
   c) Any three of the following five:
      - temperature
      - light
      - amount of water
      - volume of fertiliser applied to A and B
      - concentration of fertiliser applied to A and B.
3. Soil will already have some minerals in/is a difficult variable to control. *Or:* The method ensures that the only source of minerals is the fertiliser.
4. *Either*: Group A, which increased growth by 34%. *Or*: Group A, which had an average mass that was 50.82% more than Group B.
5. The water content of individual plants varies greatly. Dry weight measures the amount of organic material present.

   Dry weight is a measure of the amount of photosynthesis – how much sugar, starch, cellulose, protein and so on – that the plant has made. It's a much more accurate measurement of plant growth than wet weight.

6. C was the control group – to show how the plants would grow without adding any fertiliser at all.
7. There may have been a supply of mineral ions in the seed itself.
8. **QWC** Your answer should include the following points:
   - The claim is true for some species in the trial but not all.
   - Worked calculations involving named species.
   - The trial was only carried out once/The trial should be carried out more than once.
   - The trial should be carried out with a greater variety of species.

   A low level answer would include two points.
   A medium level answer would include three points.
   A high level answer would include all four points, including the worked calculations.

A3 How clean is your river?

1. 10^{-1} (one tenth or 10%).
2. 10^{-5} (one hundred thousandth, or 0.001%).
3. $16 \times 100\,000 = 1\,600\,000$ (1.6 million).
   Explanation. There were 16 individual bacteria in a 10^{-5} dilution so to get the population in the original sample we multiply it by 10^5.
4. Only the living bacteria would reproduce and produce a visible colony.
5. A standardised procedure.
6. Milligrams of oxygen per litre.
7. To prevent any photosynthesis from bacteria and algae, which produce oxygen and affect the results.
8. a) 6.1 mg O_2 l^{-1}.
   b) Moderately polluted.
9. cBOD = BOD – nBOD.
10. **QWC** Your answer should include the following points:
    - Fertiliser or sewage gets into waterways.
    - This causes high levels of nitrate and phosphate.
    - Bacterial action releases more nitrate and phosphate.
    - This causes an algal bloom.
    - Algae block the light/out-compete other aquatic plants.
    - Algae die.
    - As a result, bacterial decomposition increases.
    - Bacterial action takes all the oxygen from the water.
    - Many organisms die due to lack of oxygen.

    A low level answer would include two points.
    A medium level answer would include four points.

    A high level answer would include more than six points.

A4 The discovery of viruses

1. A testable idea.
2. There are three possible hypotheses.
   - Disease can be caused by a toxin.
   - Disease can be caused by a particle smaller than bacteria.
   - Disease can be cause by a self-replicating fluid.
3. The fluid that remains after a mixture has been filtered.
4. Viruses need to reproduce inside living cells (agar is just nutrient jelly).
5. A fine filter that can remove bacteria.
6. 6 m, 66 cm, 846 mm, 2000 µm, 226 nm.
7. a) 55 nm = 5.5×10^{-8} m.
   b) 55 mm = 5.5×10^{-2} m.
   c) 55 µm = 5.5×10^{-5}.
8. a) 0.0035 mm = 3.5 µm.
   b) 3500 nm = 3.5 µm.
   c) 4655 µm = 4.655 mm.
   d) 0.000065 mm = 65.0 nm.
9. Magnification is the amount the image has been enlarged. Resolution is the amount of detail that is revealed.
10. a) The resolving power of a light microscope is limited by the wavelength of light, which is larger than the wavelength of a beam of electrons.
    b) Any objects closer together than 5 nm will be seen as one object.
11. a) Actual size = $\dfrac{\text{observed size}}{\text{magnification}}$
    b) Observed size = actual size × magnification
    c) Magnification = $\dfrac{\text{observed size}}{\text{actual size}}$
12. a) The precise answer will depend on the size of the image on the page. You need to measure the mitochondrion with a ruler.

    The formula you need is:

    Actual size = $\dfrac{\text{observed size}}{\text{magnification}}$

    Observed size = *(for example)* 50 mm.
    50 mm is 50 000 µm.
    But the image has been enlarged 20 000 times.
    So its actual size is 50 000 µm/20 000, which is 2.5 µm.
    She/He will be out by a factor of 10.

13. a) An assertion is a statement made without evidence to back it up. In other words, it's just an opinion rather than a fact.

   b) **QWC** To answer a longer question that asks you to evaluate, the key elements are:
   - You need to give points for and against. Evaluate means *weigh up*.
   - You need to make at least three points for and one against, or one for and three against, or two and two – but not four and none.
   - You need to use continuous prose.
   - You need to use good grammar and syntax.
   - You need to use appropriate advanced language.

   Here are some points you could include:
   - It is impossible to quantify "important" in science.
   - But, if you had to do so, in terms of lives saved the microscope must have good claim to be one of the most significant inventions ever.
   - Its development has led to the discovery of cells, organelles and microorganisms, including bacteria.
   - This has allowed us to discover the causes of many diseases and therefore to prevent, by means of clean water and vaccines, and to cure, by means of antibiotics and antiseptics.
   - Other objections to the assertion include the fact that the microscope was not one invention but a series of improvements that produced greater magnification and resolution.
   - For example, it developed from Leeuwenhoek's early prototype through to the sophisticated scanning tunnelling electron microscopes that have a resolution capable of showing individual atoms.
   - Other vital breakthroughs in science include vaccines, antibiotics, the microprocessor and a dozen other developments that have had key roles in the development of the modern world.

   A low level answer would probably not acknowledge that it's impossible to quantify "important" and would evaluate the assertion in a superficial way, making only one or two points.

   A medium level answer would consider both sides but might not highlight how the microscope's not just one invention.

   A high level answer would outline at the outset that the assertion is invalid and then point out the arguments for and against before repeating that it is impossible to quantify the importance of an invention.

A5 How is diabetes diagnosed?

1. a) 10^{-3} mole, 0.001 or one thousandth.
   b) Per litre.
   c) 0.9 g (180 g in a mole, so 0.18 g in a millimole, × 5 = 0.9).

2. To establish a baseline.

   A better answer would be: So that blood glucose is not raised by recently eaten food.

3. *Either*: It's already made up and standardised/has the same composition.

   Or: It's widely available.

4. **QWC** The following points are all of advanced standard, using appropriate terminology.

   A low level answer might only include these points:
   - The levels of blood glucose are detected by the β cells in the islets of Langerhans of the pancreas.
   - If levels are high, the β cells release insulin.

   A medium level answer might also include these points:
   - Insulin is a hormone that travels in the blood.
   - Insulin fits into specific insulin receptors.

   A high level answer would include these points, as well as the low and medium level points:
   - This causes extra glucose transport proteins to be added to the membrane. (The glucose transport proteins are stored in vesicles in the cytoplasm, just under the surface membrane.)
   - This allows glucose to leave the blood and enter the cells.

   Additional points you might make:
   - The body maintains blood glucose levels within limits by facilitated diffusion. (Diffusion speeded up by specific membrane proteins is, by definition, facilitated diffusion.)
   - Insulin also activates the enzyme pathways that convert glucose into glycogen or lipid.
   - Insulin stimulates the process of glycogenesis.

   Table 29 gives examples of statements that, although correct and relevant, would not score marks.

Statement	Why it would score no marks
Blood glucose levels are detected by the pancreas.	You should know exactly which cells are sensitive to blood glucose.
Insulin causes cells to take up glucose.	You know the mechanism.
Insulin converts glucose to glycogen and lipid.	Glucose indirectly causes glucose to be converted into glycogen.

△ Table 29 Answer hints.

A6 How quickly can your body cope with sugar?

1. It's a reference value, a comparison with glucose.
2. a) (58 x 150)/100 = 87
   b) (99 x 120)/100 = 118.8
3. **QWC** The key differences between monosaccharides, disaccharides and polysaccharides are:
   - Monosaccharides are single sugars.
     Examples include glucose, fructose and galactose.
   - Disaccharides are composed of two monosaccharides joined by a glycosidic bond.
     Examples include maltose, sucrose and lactose.
   - Monosaccharides and disaccharides are classed as sugars.
   - Sugars are soluble.
   - Polysaccharides are polymers of glucose.
     Examples include starch, glycogen and cellulose.
   - Polysaccharides are large molecules.
   - They are insoluble.

   A low level answer would include up to three points.
   A medium level answer would include around five points.
   A high level answer would include seven or above points.

A7 Parasites

Table 30 gives the answers to questions 1, 2, and 3.

Weight of mouse (g)	Number of fleas	Rank of data set 1	Rank of data set 2	d	d^2
15.3	12	10	10	0	0
18.0	18	6	4.5	1.5	2.25
21.5	22	1	1	0	0
20.2	19	2	3	−1	1
16.7	14	9	8	1	1
17.9	15	7	7	0	0
19.8	17	3	6	−3	9
19.0	18	4	4.5	0.5	0.25
18.8	20	5	2	3	9
17.2	13	8	9	−1	1

△ Table 30 Answers to Q 1, 2 and 3.

4. It gets rid of negative numbers.
5. n is 10 and $\sum d^2 = 23.5$

   so $r_s = 1 - \dfrac{6 \times 23.5}{10(100-1)}$

   $= 1 - \dfrac{141}{10 \times 99}$

   $= 1 - 0.142$ so $r_s = +0.858$

6. $n = 10$.
7. Our result is significant at the 0.01 level, which is 1%.

   Either: The probability that our results are significant is over 99%.

   Or: There is less than a 1% probability that our results are due to chance.

   We can reject the null hypothesis.

8. **QWC** Your answer should include the following points:
   - In a mutualistic relationship, both species benefit.
   - Examples include coral, lichens, mycorrhizae, ruminants.
   - Many mutualisms include a producer and a consumer.
   - For example, a lichen is composed of algae inside a fungus. It is thought that the algae photosynthesises and provides organic molecules for the fungus. The fungus provides protection (from predation), moisture, carbon dioxide and inorganic ions for the algae.
   - A coral is composed of algae inside a coral animal (a polyp). The algae photosynthesises and provides organic molecules for the coral. The coral provides protection (from predation) and possible carbon dioxide and inorganic ions for the algae. By secreting a calcium "shell" the coral also holds the algae near to the sunlight.
   - In parasitic relationships one species benefits and the other suffers. For example, a tapeworm gets nutrition and protection from predation and the host gets less nutrition from the food it ingests.

   A low level answer would define both relationships but would leave out a detailed example for each one.
   A medium level answer would define both relationships and would perhaps include a detailed example for one of the relationships.
   A high level answer would define both relationships and also give detailed examples for each.

A8 Pedigree cats

1. Allele frequency will not change from one generation to the next.
2. Pedigree means *of known ancestry*. We know the family history/family tree.
3. No – their breeding is controlled by humans. It is non-random.

4. 0.15. The frequencies of all the alleles of a different gene must add up to 1 (that is, the whole population).

5. **a)** The gene pool is the sum total of the alleles circulating in a population.

   b) Pedigree breeding promotes homozygosity/reduces heterozygosity, so the gene pool becomes smaller.

6. **QWC** Your answer should include points along the following lines:
   - It promotes inbreeding.
   - Closely related individuals are bred together.
   - Individuals more likely to inherit two harmful alleles.
   - They will then be expressed.

   For example:
   - small skulls and heart problems in King Charles spaniels
   - hip problems in German Shepherds
   - kinked tails and squints in Siamese cats
   - breathing problems in bulldogs and other breeds with a shortened muzzle.

   A low level answer would just include the basic idea about inbreeding and possibly give one example.

   A medium level answer would include the fact that harmful alleles are more likely to be paired up and would include an example.

   A high level answer would clearly explain the consequences of having two harmful alleles, including the absence of the healthy allele and the subsequent genetic defect. Suitable A-level language would include the words *allele*, *homozygote* and *expressed*.

A9 Who's hiding a recessive allele?

1. hh

2. 1000. Each individual has two.

3. 80 are hh.
   80 out of 500 is 16%, or 0.16 as a decimal.
   The vital step is to realise that $q^2 = 0.16$.
   So $q = \sqrt{0.16} = 0.4$
   If q is 0.4, $p = 0.6$ (because $p + q = 1$)
   Using the equation $p^2 + 2pq + q^2 = 1$
   the frequency of the Bb genotype is $2pq$
   which is $2 \times 0.6 \times 0.4 = 0.48$
   So 48% of the population are Bb (and 36% must be BB)
   48% of 500 is 240.

4. The frequency of the allele will decrease (or, it will die out) because that genotype is a disadvantage. Predation is more likely.

5. **QWC** Your answer should include the following points:

   The Hardy-Weinberg principle states that allele frequencies will not change from one generation to the next so long as the following conditions are met:
   - The organisms are diploid.
   - Reproduction is sexual.
   - There is no migration to or from the population.
   - The population is large. The smaller the population, the greater the influence of chance. This is called *genetic drift*.
   - All genotypes are equally fertile.
   - Allele frequencies are equal in both sexes.
   - The alleles in question do not mutate.
   - Natural selection is not occurring. If it does, some genotypes will be favoured and allele frequencies will change.

   A low level answer would include any three points from the list.

   A medium level answer would include any five points from the list.

   A high level answer would include seven to eight points from the list.

A10 Pesticides, pollutants and food chains

1. 10^9.

2. 0.02 ppb in standard form is 2×10^{-11}.

3. 20 000 ppb is 20 ppm.

4. 100 000 times (20 000/0.2).

5. 25 ppb (100 after 72 days, 50 after 144 days).

6. **a)** Modelling or computer modelling.

   b) • Ecosystems are very complex.
   - There are many variables that cannot be predicted; for example, changes in climate/temperature/rainfall.

7. **QWC** The following points could be used in your answer:
   - It's true that most extinctions have nothing to do with man.
   - However, this doesn't mean we have no responsibility.
   - Conservation is important for a variety of reasons.
   - Healthy ecosystems: ecosystems are balanced and the removal of any species can upset the balance.
   - Conservation maintains genetic diversity.
   - It's not fair to future generations.
   - Removal of one species due to human activity may shift the balance of the ecosystem and lead to the extinction of another species.

- Some species might provide useful products, such as drugs.
- Any suitable example, e.g. antibiotics from fungi, anaesthetics from tree bark.

A low level answer would include three points from the above.

A medium level answer would include five points.

A high level answer would include seven points and an example.

A11 Can we control an animal's growth rate?

1. Exactly the same in all respects, except that they are given an inactive injection, e.g. saline (a harmless weak salt solution) instead of thyroxine.

   NB It is important that all the rats got the same treatment and handling, so that the only difference was the hormone. Otherwise, you could argue that the stress of handling and being given an injection could have affected the growth rate.

2. Thyroxine has no effect on the growth of rats.
3. There is a low probability that the results are due to chance.
4. Randomly, so the two groups are as identical as possible. Any from: genetically similar, same age, same diet, same sex or balance of sexes.
5. Add up all the values; divide by the number of values.
6. A large standard deviation = large spread of data (many values far from the mean). A small standard deviation = small spread of data (most values close to the mean).
7. The mean value is 3.55 and 68% of the sample falls between 3.07 and 4.03.
8. Experimental group = 0.48/3.32 = 0.14; Control group = 0.22/3.46 = 0.06
9. a) The purpose of journals is to publish work so that others in the field can see what research has been carried out. Research published in journals adds to the body of human knowledge, means scientists can avoid wasting their time by repeating research that's already been done, and provides ideas for new research and new hypotheses to explore.

   b) Peer review allows errors/faults/bad science to be spotted by experts in similar fields of research.

   c) **QWC** Remember that *evaluate* means you need to weigh up both sides. Possible points include:

   For:
   - The results were significant.
   - The probability that the results are due to chance is low.
   - Thyroxine does affect growth rate.

   Against:
   - This is just one species of rodent.
   - Other species might respond differently.
   - There is far too small a sample size to draw valid conclusions.
   - The investigation was only carried out once – it needs repeats.
   - The investigation was only done for four weeks, not over the animals' whole growth phase.

   A low level answer would generally include one point for and two against.

   A medium level answer would include two points for and three against.

   A high level answer would include three points for and four against.

A12 Who's at risk of heart disease?

1. < means less than.
2. a) They have inherited genes/alleles that make them more likely to develop heart disease.
   b) *Either:* Because a change in lifestyle will not affect the genes they have inherited. They cannot do anything about it.

   Or: It is expensive/difficult to analyse an individual's genes.
3. Any three from:
   - The higher the cholesterol, the higher the risk.
   - The higher the systolic blood pressure, the higher the risk.
   - Smoking increases the risk. (Or: The more cigarettes smoked, the higher the risk.)
   - The higher the HDL level, the lower the risk.
4. 26% of 600 000 is 156 000.

   Work it out as $\frac{600\,000}{100} \times 26$

   (Think: 1% of 600 000 is 6000; 26 × 6000 = 156 000.)
5. a) His points score is 15 (7 for being in his 50s, 3 for being a smoker, 4 for his cholesterol, 1 for his blood pressure and 0 for his HDL).

   From Table 19, that's a 20% increase in his chances of having a heart attack.

   b) **QWC** Acceptable points include:
   - Stop smoking.
   - Reduce high cholesterol foods; for example, animal fats, dairy products, eggs.
   - Reduce blood pressure; for example, reduce salt in the diet, take more exercise.
   - Lose weight.
   - Gentle exercise such as walking or swimming is better than vigorous exercise.

- Take medication such as statins to lower cholesterol, or ACE inhibitors to reduce blood pressure.

A low level answer would include two or three general points.

A medium level answer would include four points including more detail, such as examples of high cholesterol foods and ways to reduce blood pressure.

A high level answer would include six points including specific examples of high cholesterol foods and ways to reduce blood pressure.

A13 The potential of stem cells

1. An undifferentiated cell that can replicate indefinitely in its undifferentiated form, and also give rise to various differentiated cell types.

2. Potency is the ability of a stem cell to differentiate into different specialised cells. The more potent the stem cell type, the greater the variety of specialised cells it can become.

3. - Totipotent stem cells (from the early embryo) can differentiate into any cell type and can become a whole organism.
   - Pluripotent stem cells (from the inner cell mass of a blastula) can differentiate into any cell type but cannot become a whole organism.
   - Multipotent stem cells (from adult bone marrow tissue, for example) can differentiate into a few, closely related cell types.

4. Any three from (this list is not exhaustive):
   - Alzheimer's disease
   - diabetes (Type 1)
   - burns (new skin)
   - heart disease
   - spinal cord injury.

5. **QWC** Remember you need to weigh up both sides in order to evaluate the statement. Full marks given for 3/1, 2/2 or 1/3 split of points.

For (max. 3 points):
- Embryonic stem cells have greater potential to turn into more cell types.
- They could therefore be used to treat a wide variety of illnesses.
- Give the example of a disease/condition that could be treated.
- The embryo would only be thrown away anyway.
- It's just a few cells, not a conscious/thinking individual with a brain or an identity.
- The people who donate an embryo do so with their knowledge/permission that it may be used for stem cell research.
- Regulated repositories of embryonic stem cells are now available as a resource for research, so it isn't always necessary to use new embryos.

Against (max. 3 points):
- An embryo is a human life and should be given rights.
- Adult tissues can also provide effective stem cells.
- It's a slippery slope. Continuing work with embryonic stem cells may lead to unacceptable practice, such as attempts at human cloning.
- People may feel pressured into donating embryos, or expect to receive payment.
- New technology means that new embryos are rarely needed.

A low level answer would include one point for and one against.

A medium level answer would include two for and two against.

A high level answer would have at least two for and two against or three for and three against, with an attempt to justify the points.

NB "We should not play God" is never a good answer. What does this mean? If it means that we should not interfere in an individual's health and just let nature take its course, we have been "playing God" since medicine began. Saying "It's not natural" is also a meaningless answer.

A14 Where have all the UK's forests gone?

1. a) Mowing reduces plant species diversity.
   b) Mowing has no effect on species diversity.

2. Taking a few measurements in the hope that they will be representative of the whole (or words to that effect).

3. a) *Either*: Without bias. *Or*: Without the conscious choice of the experimenter.
   b) Map out the two areas into squares. There are various ways to do this, such as using pins and string, or photographing the area and drawing gridlines.

   Give each square a number.

   Select which numbers to put quadrats in:
   - by generating random numbers
   - by using a book of random numbers (yes, they do exist), a telephone directory or, more realistically, a computer or calculator with a random number function.

4. - Numbers of different species.
   - Numbers of different individuals of different species.

5. This investigation does not need a control. It's a comparison of two areas. Each area acts as a control for the other.

6. **a)** Some species would be missed.

    b) No more species would be found, so including more quadrats would just waste time.

7. Between 15 and 17.

8. **QWC** The answer to this question is in the QWC Worked Examples section.

A15 Overweight or just big-boned?

1. 17.1, 18.1, 19.2, 22.0, 22.3, 23.6, **24.4,** 24.6, 24.7, 24.7, 24.7, 25.4, 26.4, 33.0

2. **a)** mean = 330.2/14 = 23.59

    b) median = 24.4

    c) mode = 24.7

    Note: All of these values are comparative. They show the position of your value compared to the whole population/sample/cohort.

3. **a)** Percentile – position to within 1%.

    b) Decile – position of a value in tenths, for example top 10%.

    c) Quartile – position of a value in quarter blocks of 25%. The interquartile range is the middle 50% of the sample: 25% above and below the 50th percentile.

4. **a)** Systematic error.

    b) This sort of error can be eliminated by calibrating scales. Use known weights to check accuracy.

    c) Inaccurate – all readings are 5% below what they should be.

    The data is precise. It is in suitable units and allows us to measure what we want to measure. There would be no point in measuring in grams, which would give us different readings every time. If we repeated the measurement on the same scales they would give very similar values.

5. **a)** It's norm-referenced. It compares a child's progress to others in the same country.

    b) By taking measurement of many children over many years.

6. **a)** The heaviest child weighs 33 kg.
    - On the 95th percentile.
    - Only 5% of children are heavier at this age.

    b) The lightest child weighs 17.1 kg.
    - Below the 5th percentile.
    - Over 95% of children are heavier at this age.

7. 21.5–27kg (allow 0.5 kg either way).

8. **a)** BMI = 26.6.

    b) Overweight.

9. Scatter graph/scattergram.

10. A positive correlation.

11. **a)** Any individual in the lower-right quadrant/portion of graph.

    b) A heavily muscled individual, perhaps a sportsman/athlete. (*NB*: do not say "heavy bones".)

12. **QWC** Remember that you need to weigh up both sides when evaluating.

    For:
    - It would be very useful for predicting risk.
    - Early diagnosis of disease might improve chances of survival..
    - Likely diseases are diabetes, coronary heart disease and lung cancer.
    - It would allow resources/specialist staff to be brought in.
    - Some basic information is already available, for example, age and gender.
    - It would encourage people to improve their lifestyle.

    Against:
    - The rest of the information would be very time consuming to gather.
    - People have to volunteer the information – it can't be compulsory.
    - This means not everybody would participate.
    - People lie/underestimate, especially about alcohol, smoking and diet.
    - It would involve gathering information about a wide range of variables for each individual.
    - Gathering one-time information on factors such as blood pressure and cholesterol levels may not be sufficient.

    A low level answer would include at least three points: two for and one against or vice versa.

    A medium level answer would include any five points with a three/two or two/three split.

    A high level answer would include any seven points with a three/four or four/three split.

AIS1 Investigating the effect of temperature on the activity of trypsin

1. Temperature

2. Rate of reaction/time taken to digest the egg white.

3. Any three from:
    - size of glass tube/surface area of protein
    - length of tube/amount of protein
    - pH
    - volume of enzyme
    - concentration of enzyme.

4. Repeat the experiment with no enzyme/boiled enzyme, to show that it was the enzyme that was digesting the protein and not another factor, such as temperature.

5. If the temperature were any lower the rate of reaction would be very slow. If it were any higher the enzyme would be denatured.
6. It needs to show two sets of continuous data.
7. High numbers indicate a higher rate (as opposed to time, where low numbers indicate a high rate).
8. To make the numbers more familiar/easier to work with.
9. a) Interpolation.
   b) 3.4.
10. a) Extrapolation.
    b) 0.
11. No, this is not a valid conclusion. The enzyme activity at temperatures around the optimum – 38 or 42, for example – may be higher. These temperatures should be investigated too.
12.
$$Q_{10} = \frac{\text{rate at 30}}{\text{rate at 20}}$$
$$Q_{10} = 4.5 = \frac{4.5}{2.4} = 1.875$$

    Note: Most Q_{10} values are around 2, which means that the rate of reaction doubles for every 10 °C rise. Enzymes are very temperature sensitive.
13. • Enzymes are globular proteins.
    • They have a precise tertiary structure.
    • They are held together with hydrogen bonds and other weak forces.
    • A precise area on the surface – the active site – acts as the catalytic centre.
    • The substrate and active sites are complementary.
    • The molecular shape and chemical charges attract each other.
    • Enzymes are specific because only one substrate matches the active site.
    • Enzymes are denatured by higher temperatures because the tertiary structure changes.
    • This happens because the molecule vibrates so much that the hydrogen bonds/weak bonds are broken.
    • pH also affects the shape of the active site.

    A low level answer would include any three points.
    A medium level answer would include any five points and would use scientific language.
    A high level answer would include any seven points with the correct use of scientific language.

AIS2 Osmosis investigations

The basic idea is that the potato cylinders in solutions with a higher water potential will absorb water and gain weight. Those in solutions with a lower water potential will lose water, so weigh less. We can use a graph of weight change against concentration to estimate the water potential of the cytoplasm in the cells.

1. Omsosis is the net movement of water from a region of high water potential to a region of lower water potential across a partially permeable membrane.
2. Osmosis will happen more rapidly in salt than in sucrose, because there are twice as many particles to attract water.
3. There are eight missing values (Table 31). Score 1 mark for each correct row of 4 (2 marks max.).

Concentration (M)	0.0	0.1	0.2	0.3	0.4	0.5
ml salt or sugar	0	4	8	12	16	20
ml water	20	16	12	8	4	0

△ Table 31 Answer to Q 3.

4. a) The independent variable is the concentration of the solution.
   b) The dependent variable is the weight change in the potato, which is a measure of how much osmosis has taken place.
   c) Controlled variables – any three from:
      • the size of the piece of potato
      • the surface area of the piece of potato
      • the composition of the potato (all pieces should be from the same potato)
      • the temperature
      • the length of time they were kept in the solutions.
   d) Factors that could give inaccurate results – any two from:
      • The potato cells/tissue could be different in different parts of the potato. For example, the water potential of the cells at the surface could be lower due to evaporation from the potato surface.
      • The balance could be inaccurate.
      • Human error in making the solutions.
      • Variations in the surface area of the potato.
      • The skin could have been left on a piece of potato thus lowering the surface area for osmosis.
5. Osmosis stops when an equilibrium is reached. Cells either become turgid or lose water until the water potentials are equal inside and outside the cell.
6. Any two from:
   • It avoids introducing a systematic error.
   • The balance may be inaccurate.
   • The error will be the same for all results.

Concentration (moles)	Initial weight (g)	Final weight (g)	Difference (g)	% Change in weight $\frac{\text{Change in mass}}{\text{Original mass}} \times 100$
0.0	9.58	10.27	0.69	+7.2
0.1	9.72	10.03	0.31	+3.2
0.2	9.35	9.29	−0.06	−0.6
0.3	9.90	9.32	−0.58	−5.9
0.4	9.62	8.80	−0.82	−8.5
0.5	9.81	8.57	−1.24	−12.6

△ Table 32 Sample results: salt

Concentration (moles)	Initial weight (g)	Final weight(g)	Difference (g)	% Change in weight $\frac{\text{Change in mass}}{\text{Original mass}} \times 100$
0.0	10.13	11.04	0.91	+8.9
0.1	8.60	9.09	0.49	+5.6
0.2	8.89	9.19	0.30	+3.3
0.3	9.77	9.85	0.06	+0.6
0.4	9.27	9.09	−0.18	−1.9
0.5	9.19	8.75	−0.44	−4.7

△ Table 33 Sample results: sucrose

7. The missing values are given in Tables 32 and 33. Score 1 mark for each correct column (2 for salt, 2 for sucrose, max. 4 marks).
8. a) The 0.0 M (that is, pure water) solution has the highest water potential.
   b) The 0.5 M solution has the lowest water potential.
9. Percentages allow comparisons between differently sized pieces of potato.
10. Increase the temperature; increase the surface area of the potato pieces.
11. a) No change/no osmosis.
    b) Potato cylinder lost water/weight.
    c) Potato cylinder gained water/weight.
12. Isotonic means equal water potential (between two solutions, cells or tissues).
13. a) For salt, any value between 0.18 and 0.19 M.
    b) For sucrose, 0.34 M.
14. Using the curve for salt, the water potential of the potato tissue is about −1000 kPa.

    Using the curve for sucrose, the value is about −950 kPa.

    Ideally, both values should be the same.
15. Points you could consider in your answer include:
    - Water is a small molecule, Rmm 18. Most molecules that size are gases. But water is a *polar molecule* – it has an area of positive and negative charge, so it's a bit like a mini magnet. It is attracted to other water molecules, and to any other polar substance, because hydrogen bonds form between areas of positive and negative charge. It is this key feature that is responsible for water's unique properties.
    - Water is a liquid over most of the temperature zones encountered on Earth.
    - It is a solvent for ionic/polar/charged substances.
    - Water's cohesive nature creates surface tension. It also forms long, unbroken columns in xylem vessels that conduct water to the top of tall trees.
    - Water has a high latent heat of evaporation. It produces a cooling effect as it evaporates.
    - Water is a thermal buffer. It can gain or lose a lot of energy before the temperature changes.

AIS3 Using the chi-squared test

1. The red squirrel prefers to nest in a particular species of tree.
2. The red squirrel has no preference for the type of tree in which it nests.
3. 17.5 in all the boxes (12 + 18 + 15 + 25, divided by 4).

4. The missing values are given in Table 34. Score 1 mark for each correct row (max. 5 marks).

Species of tree	Larch	Spruce	Scots pine	Fir
Observed results (O)	12	18	15	25
Expected results (E)	17.5	17.5	17.5	17.5
O – E	–5.5	0.5	–2.5	7.5
(O – E)²	30.25	0.25	6.25	56.25
$\frac{(O-E)^2}{E}$	1.73	0.01	0.36	3.21

△ Table 34 Answer to Q 4.

5. The value of χ^2 is 5.31 (worked out as 1.73 + 0.01 + 0.36 + 3.21).
6. Three degrees of freedom. (There are four categories, minus one.)
7. It means that the probability that our results are due to chance is less than 5%.
8. The third row should be highlighted.
9. a) Our χ^2 value of 5.31 falls between 4.11 and 6.25.
   b) This falls between p values of 0.25 and 0.1.
10. The probability is between 25% and 10%.
11. The results are not significant.
    So either: We must accept the null hypothesis.
    Or: The squirrels are probably not expressing a preference.

QWC Worked Examples

At some point during your course, you will be assessed on the quality of your written communication. These annotated worked examples show how low, medium and high level responses to Question 8, Activity 14 gain the marks they do.

8. Energy is transferred through an ecosystem. Describe how and explain why the efficiency of energy transfer is different at different stages in the transfer. **QWC** S18–22, 39

★ Low level answer

This is a very weak answer. It consists of loose language and contains very little that the candidate has learned at advanced level. Also, it doesn't contain enough points. If this were a six mark question, to actually score six marks, the student would need to give the examiner six points of advanced standard material. There are two command words in the question, "describe" and "explain", so you need to give reasons rather than a list of simple facts.

Specific problems:

Energy cannot be created or destroyed. It is *captured* in photosynthesis.

Photosynthesis is spelled incorrectly.

Plants are not the only producers; algae and some bacteria can photosynthesise too.

Dung isn't a very scientific word, although there are worse. Faeces or faecal matter is better.

Ecosystems do not just consist of animals eating grass. The candidate has latched onto one example they have been taught.

There's no evidence of understanding *why* energy is lost, and why it is different at different stages.

> *Energy is made by photosinthersis. Producers are plants. Consumers eat the producers. Energy is lost at each level because when animals eat the grass lots of it can't be digested and so is lost in the dung. Eventually all the energy runs out, which is why animals at the top of the food chains are rare.*

★★ Medium level answer

This is a better answer. There is some attempt at a structured answer and the language is more appropriate to advanced level.

In this answer:

- Arguably the first point isn't part of the question, but it's neutral as far as the examiner is concerned and does show that the candidate is structuring the question logically.

- The second point is perfectly valid but should contain more detail about why photosynthesis is inefficient and where the energy is lost.

- The third point is valid but again lacks detail.

- The fourth point is too simplistic – and it should be "bodies", not "body's". It would be much better to say that energy is contained in the chemical bonds of the organic molecules that make up the animal's body.

- The fifth sentence is focused on the question, but lacks detail. What does "earlier" and "later" mean, and why are they different?

- The last point is a good one and is worthy of a mark, but it misses the point that *respiration* is the inefficient process that produces the heat. And animals respire even if they aren't moving around.

> *All energy comes from the sun. Energy is captured by plants in photosynthesis. Animals eat the plants and the energy is transferred to them. The energy passes up the food chain in the body's of the animals. Later transfers are more efficient than earlier ones. Eventually all the energy is lost as heat, because the animals use up energy as they move around.*

★ ★ ★ High level answer

This is a top level answer because:

It is logically structured as it begins with the entry of energy into the ecosystem, then covers the transfer of energy from producer to consumer, and finally from consumer to consumer. It has paid attention to the question's command words.

It uses appropriate language throughout. such as "photosynthesis" and "chlorophyll".

It shows understanding by explaining the *processes* that are involved in energy transfer and energy loss.

It is also quantitative, giving approximate energy transfer values.

Energy enters the ecosystem by photosynthesis. Energy is lost because some sunlight energy misses chlorophyll, some is reflected off the plant and some is not the right wavelength. Generally, red and blue light is absorbed while green is reflected. This sun [ensure next figure is a right facing arrow] --> plant transfer is rarely more than 2% efficient.

Efficiency of transfer from producer to consumers is approximately 5-10%; all organisms lose energy as a consequence of respiration. Energy from respiration is lost to the environment as heat.

Energy transfer from producers to herbivores is inefficient because there is so much cellulose, which cannot easily be digested.

Energy transfer from consumer to consumer is more efficient because meat is more easily digested. Efficiency is lower in warm-blooded animals because they expend more energy maintaining a constant body temperature.

Glossary

abstract A summary at the start of a scientific paper that outlines what was investigated, by what method, and what the findings were. Very helpful when deciding if a paper is worth reading.

accuracy How close measurements are to their true value.

anomaly A result or piece of data that does not fall within the normal range. This might be due to human error in measuring or calculating, or it may be an exceptionally large or small specimen. When plotting data on a graph, anomalous data will be furthest from the line of best fit.

arbitrary units Units that have been made up just to keep things simple. Don't worry about the units. It's the comparison between the values that is important.

argument Not to be confused with aggression or a fight. Arguments in science are a healthy and important part of the process, as people discuss opposing standpoints. "I think this, because…".

assertion A statement that is opinion rather than fact, because it is not backed up by evidence.

bar chart Type of graph that illustrates discrete categories, such as blood groups. Different from a histogram.

base unit The basic SI units of measurement: metre, kilogram, second and mole are the commonly used base units in biology.

bias Conscious choice by the experimenter. Statistical tests can only be done on data gathered *at random*, which by definition means *without bias*.

blind trial Trial of drug or other treatment in which the participants don't know whether they have been given an active or an inactive dose. Its purpose is to overcome the placebo effect.

bod Biochemical Oxygen Demand. A measure of water quality by measuring the oxygen content of a sample before and after incubation. The more bacteria and organic waste there is in a sample, the more oxygen will be taken out of the water.

carcinogenic Cancer-causing. Carcinogens include ionising radiation, cigarette smoke and various chemicals.

causal relationship Where one factor causes another. For example, the more intense the sunshine, the higher the incidence of skin cancer. Compare with correlation.

chi-squared A statistical test that compares the observed results with those you would expect if the results were purely due to chance. Suitable when the data fits into categories.

clinical trial Part of the drug development process that involves testing on humans.

colorimeter Machine that measures the absorbance, and therefore the transmission, of light through a solution. If 100% of light is transmitted, 0% is absorbed.

conclusion A summing-up of the key finding of an investigation. It relates to the original hypothesis and ideally should be one sentence.

confidence limits An extension of standard deviation. The error bars above and below the mean show the range in which you would be 95% confident that any further values will fall.

continuous data The opposite of discrete data. Continuous data shows a range of values, typically with most values in the middle. See "normal distribution".

continuous variation Variation that shows a range of values.

control A comparison. A repeated investigation which is the same apart from the variable being investigated. This validates the results by showing they are due to the factor being studied and nothing else.

control group In trials, the control group should be an equivalent group that is treated in exactly the same way as the experimental group but not given the active treatment/drug. See placebo and placebo effect.

çcorrelation A match in the pattern of data. Not to be confused with a cause.

counter argument The opposing point of view.

data Information, which can be qualitative or quantitative. See also primary and secondary data.

datum A piece of data. Data is plural, datum is singular.

decile Tenth. See also percentage, quartile.

decimal Simply, a way of displaying a value, as opposed to a percentage, ratio or probability.

dependent variable In an investigation, the *effect*. The factor that is measured.

derived units A unit that is a multiple or a product of one or more of the SI base units. Examples include m² for area, m³ for volume and kg per cubic metre (kg m^{-3}) for mass density.

discrete data Data that falls into categories. Also called discontinuous data. For example, blood group can be A, B, AB or O, but nothing in between.

diversity index A measurement of biodiversity that results from the number of different species and the number of individuals of each species. The two most commonly used are Simpson's Diversity Index, formula

$$D = \frac{\Sigma n(n-1)}{N(N-1)}$$

And Simpson's Reciprocal Diversity Index,

$$D = \frac{N(N-1)}{\Sigma n(n-1)}$$

Where N is the total number of organisms of all species and n is the total number of organisms of each species, and Σ (sigma) means the sum of all.

double blind Trial of drug or other treatment in which neither the participant nor the person administering the trial know who is getting the active treatment and who isn't. Designed to eliminate the placebo effect.

error bar A way of displaying standard deviation on a graph. Error bars show the spread of the data about the mean. Often used as a visual way of seeing if two sets of data are significantly different from each other.

ethics Area of science concerned with morals. What *should* we do?

extrapolation When using a graph, predicting values by continuing the line. For instance, using a graph of human population, we can predict what the world's population will be in 2020 by extrapolation.

fair test An investigation in which all variables have been controlled apart from the two under investigation, allowing you to conclude that the effect was due to the cause.

frequency diagram A type of histogram or bar chart in which the *y* axis is frequency. Used to show the distribution of different categories or values.

gradient The steepness of the graph. Change in *x* divided by change in *y*. Can be used to show enzyme activity, for example, or rate of population growth.

histogram Graph in which continuous numerical data is divided into categories, and the bars have no gaps between them.

human error Inaccuracies in collecting data due to the limitations of the experimenter. This does not mean that the experimenter is incompetent. The limitations of our senses, and steadiness of hand, can contribute, as can simple chance.

independent variable In an investigation, the factor being varied. Simply, it's the *cause*, while the dependent variable is the *effect*.

interpolation Using a graph to predict a value that wasn't measured. Think of this as reading "between the points".

inversely proportional As one variable increases, another decreases. For example, as the distance to a light source increases, the light intensity decreases.

in vivo In living cells, or whole organisms. It literally means "in life".

in vitro In glass – test tubes and Petri dishes. Out of the body.

journal A publication that contains the latest research on a particular topic.

kick sample Ecological sampling method commonly used in moving water. Organisms under rocks are dislodged and a carried into a net held downstream.

line graph A line drawn to show the relationship between two numerical sets of data.

line of best fit Line drawn between scattered data to show underlying trend. Best done by a computer.

logarithmic scale A scale on a graph with geometric increments such as 10, 100, 1000, 10 000, and so on. It allows a large range of values to be shown on a manageable scale.

nominal data Data that has names rather than values: blood groups or socio-economic groups, for example.

normal distribution The spread of the data about the mean. Middle values are common and extremes are rare. A graph showing a normal distribution is often described as *bell shaped*.

null hypothesis The default position of an investigation. For example: "Caffeine will have no effect on a heartbeat".

open label trial Trial in which the patient and the experimenter know which treatment is being given. Contrast with blind and double-blind trials.

ordinal data Data that is ranked in order, for example from heaviest to lightest. The exact quantity of a value is often less important than its position in the overall rank.

paradigm A framework of thinking. A paradigm shift is a change in the way we think. For example, everyone thought that the Earth was the centre of the solar system, and everything revolved around us, until Copernicus suggested that the Earth, and all the other planets, revolve around the Sun.

peer review A quality control process for scientific research. Before scientists publish their work in journals, it is reviewed by peers (or *equals*) who are scientists working in the same area, because they are the ones most likely to spot any flaws in the methods, data handling or conclusions.

percentile A value below which a certain percentage of values lie. For example, if your intelligence is said to be on the 90th percentile, it means that only 10% of people are equally or more intelligent than you and 90% are less intelligent.

placebo A treatment that has no active ingredient. For example, a pill made of sugar or chalk, or an injection of water or saline (weak salt solution).

placebo effect A positive improvement in health that is not due to a particular medicine or invasive treatment, such as surgery. For example, a patient's pain is perceived as getting less after the doctor has given some ineffective medicine – a placebo – such as a sugar pill or water injection.

plagiarism The dishonest practice of passing off someone else's work or ideas as your own. To avoid plagiarism, work should be properly referenced.

primary data Data that is collected directly by the investigator. Contrast with secondary data.

prognosis In medicine, the likely outcome of a disease or condition. If you are a heavy smoker, old, obese, with high blood pressure and a family history of heart disease, the prognosis is not good.

protocol A standard set of instructions for carrying out a procedure or investigation so that results can be properly compared and replicated if necessary

punnet square In genetics, a diagram that helps to organise all possible combinations of gametes, to work out the outcome of a particular cross.

quadrat A small frame that allows the sampling of an area. Used in ecological investigations.

qualitative data Data that is non-numerical.

quantitative data Data that has numerical values, for example, height and weight.

quartile Divisions of 25%. See also percentile and decile.

random error Inaccurate data that is just as likely to be above or below the true value.

random number Numbers selected without any conscious choice.

randomisation The process of fairly allocating individuals to either an experiment or a control group, so that both groups would be expected to progress the same way without the treatment under investigation.

ratio A way of comparing two values that measure the same thing.

reagent A chemical substance used in a procedure, For example, Benedict's solution is the reagent used to test for a reducing sugar.

reciprocal 1/value. The reciprocal of 2 is ½. In investigations that measure time taken, for example, 1/time is a useful measurement of rate.

references List of sources, usually at the end of a piece of scientific writing. They give credit for other people's work.

referencing The process of listing all the sources used in a piece of work.

regression In statistics, regression analysis is a statistical process for estimating the relationships among variables.

regression curve A graph that shows how two variables are statistically related. Note that the "curve" can be a straight line.

reliability Data is reliable if the same results could be obtained by different people performing the same investigation.

reproducibility A central idea in the scientific method. Data is not reliable or valid unless it is reproducible, meaning that when the procedure is repeated, either by the experimenter or different individuals, similar results are obtained.

resolution The power of a microscope to reveal detail. A resolving power of 10 nm means that objects closer together than 10 nm will be shown as one object.

running mean In sampling, a technique that shows how many measurements need to be taken. Every time a reading is taken, a mean is calculated. When the mean stops changing significantly, you have got representative data.

sampling Taking a few measurements when it is impossible to take them all.

scattergram Graph on which two sets of data can be plotted to see if there is a correlation.

secondary data Data that is collected by someone other than the investigator/user. The advantage of secondary data is that it will usually have already been established as valid and reliable. Contrast with primary data.

serial dilution Technique for making progressively more dilute solutions, usually a factor of 1 in 10 each time.

SI International System of Units (abbreviated SI from French le Système Internationale). Units of measurement used all over the world. The standard units are metre, kilogram, litre and second. See Appendix 1.

significant figure The number of places after the decimal point. A reading like 3.41237 would be 3.41 correct to two significant figures.

skew An inaccuracy in results.

Spearman Rank Statistical test that compares two sets of ranked data to see if there is a correlation. For example, do the countries that smoke the most cigarettes per capita have the highest incidence of lung cancer?

standard deviation A measure of the spread of data about the mean. One standard deviation is the range of values that includes 68% of the sample. Two standard deviations include 95% of the sample. Usually written as x ± y.

standard error Method of using standard deviation to give confidence limits. For example, if two times the standard deviation are plotted on error bars above and below the mean, we can be 95% confident that any further readings would fall within the error bar.

standard form Way of expressing large or small numbers in the form of $A \times 10^x$, where A is a number between 1 and 10. For example, 0.00074 in standard form is 7.4×10^{-4}.

systematic error Error that is always above or below the true value by a consistent amount. Usually due to faulty equipment.

t-test Statistical test that determines if there is a significant difference between two sets of similar data. t-tests compare the mean and standard error of two sets of data to see if they are significantly different

teratogen Environmental agents or factors that cause abnormalities in the embryo/foetus. Thalidomide is one of the most famous examples of a drug that wasn't trialled for its teratogenic effects.

theory Central ideas or conceptual framework used to explain scientific observations and make predictions.

transect Sampling method that can be used to show the transition between two areas, such as down a rocky shoreline, or a sand dune. Transects involve systematic sampling. See also quadrats.

venn diagram Graphical representation that shows the degree of overlap between two or more areas.

x axis In a graph, the axis that goes along the bottom of the page, usually from left to right. In an investigation, the x axis usually contains the independent variable.

y axis In a graph, the vertical axis, which will often contain the dependent variable.

Appendix

International System of Units (SI)

SI Base Units

Base quantity	Name	Symbol
Length	metre	m
Mass	kilogram	kg
Time	second	s
Temperature	Celsius	°C
Amount of substance	mole	mol

SI Derived Units

Derived quantity	Name	Symbol	Equivalent
Force	newton	N	m kg s^{-2}
Pressure	pascal	Pa	N/m^2
Energy	joule	J	N m

SI Prefixes

Factor	Name	Symbol	Numerical value
10^{12}	tera	T	1 000 000 000 000
10^{9}	giga	G	1 000 000 000
10^{6}	mega	M	1 000 000
10^{3}	kilo	k	1000
10^{2}	hecto	h	100
10^{1}	deka	da	10
10^{-1}	deci	d	0.1
10^{-2}	centi	c	0.01
10^{-3}	milli	m	0.001
10^{-6}	micro	μ	0.000 001
10^{-9}	nano	n	0.000 000 001
10^{-12}	pico	p	0.000 000 000 001

Frequently used symbols

Symbol	Means	Example with explanation
=	equals	x = y means x is the same as y
>	is greater than	x > y means x is greater than y
<	is less than	x < y means x is less than y
>>	is a lot greater than	x >> y means x is a lot greater than y
<<	is a lot less than	x << y means x is a lot less than y
∞	is proportional to	x ∞ y means that x is proportional to y
~	is approximately equal to	x ~ y means x is approximately equal to y

Index

A
abstracts 10
accuracy 12, 13
adipose 61
algae 50, 59
allele frequencies 14, 30–1, 48, 49
animal testing 9
antibiotics 37
antibodies 29
atheroma 55
atherosclerosis 55
ATP (adenosine triphosphate) 16

B
bacteria 14, 37, 40–1
beans 45
Beirjerinck, Martinus 42
bias 13
bioaccumulation 50
bioamplification 50
biochemical oxygen demand (BOD) 40–1
bioethics 58
bioinformatics 14
biometrical impedance analysis 61
blackbirds 23
blind trials 8
blood clots 55
blood glucose 44, 45
Bloom's Taxonomy 17
body fat 61
body mass index (BMI) 62
bone marrow 57

C
Calvin cycle 16
cancer 6, 26, 27
carbohydrates 45
carbon and nitrogen cycle 16
cardiac cycle 16
cats 48
cells 9, 16, 25, 61
  cell division 16
Chamberland, Charles 42
chi-squared tests 34
chloride 37
cholera 37
cholesterol 8, 27, 28, 55
clinical trials 9, 10
clover 59
cohort studies 55
command words 17–19
communities 16
complex carbohydrates 45
computer programs 14
concentrations 24–5
conclusions 5, 12, 13
contamination of waterways 40
continuous data 26
continuous variation 32
control experiments 7
control groups 7
coronary heart disease 55
correlations 27–9, 34, 46, 47
Curie, Marie 4
Cycles in Biology 13, 15
cystic fibrosis 49

D
Darwin, Charles 4
data 10, 12
  accuracy 12, 13
  bias 13
  chi-squared tests 33
  continuous 28
  continuous variation 33
  correlation 28–9, 34
  errors 12–13
  nominal 23
  normal distribution 32
  ordinal 23
  precision 12
  qualitative 26
  quantitative 26
  reliability 12
  repeatability 12
  reproducibility 12, 13
  sampling 29
  Spearman rank tests 34, 46
  standard error 34
  statistical tests 34
  t-tests 33
  validity 12
deciduous forests 59
deciles 23
decimal places 12, 26
dehydration 37
dependent variable 5
detection-corrections 44
DEXA (dual-energy X-ray absorptiometry) 61
diabetes 44
diarrhoea 37
diploid species 49
double-blind trials 8, 37
drug development 9–10
dysentery 37

E
ecosystems 14, 50, 51, 59
egg whites 7
Einstein, Albert 4
electrolytes 37
electron microscopes 21, 42
embryos 57
endothelium 55
environmental considerations 14
enzymes 7, 48
epidemiology 55
error bars 32, 53
errors
  human 12
  random 12
  systematic 13
estimates 21
ethics 58
eutrophication 51
Excel 32
experimental hypothesis 6, 31
extra-embryonic tissue 57
extrapolation 28

F
Faraday, Michael 4
fat 61
fertilisers 38
field work 14
food chain 50–1
forests 59
fungi 59

G
genes 30, 48, 49
genotypes 31–2
glucose 37, 44, 45
glycaemic index 45
glycaemic load 45
graphs
  error bars 32
  extrapolation 28
  graph types 27–8
  histograms 27
  interpolation 28
  line graphs 28
  line of best fit 28, 29
  logarithmic scales 29

quadrats 29
reading of 28
running mean 29
sampling 29
scattergrams 28–9
grazing animals 59
guinea pigs 49

H
Hardy-Weinberg equilibrium 30–1
hazards 13–14
heart disease 55
hematopoietic stem cells 57
heterozygotes 30, 49
histograms 28
HIV (human immunodeficiency virus) 42
homeostasis 44, 45
homozygotes 31, 49
hormones 52
human error 12
hydroponics 38, 39
hypotheses 4–6, 13, 17, 32, 46
experimental 6, 32
null 6, 32

I
ideas 4–5
identical twins 7
immune response 30
in vitro fertilisation (IVF) 57
in vitro testing 9
independent variable 5
infection risk 14
insulin 45
interpolation 29
ions 37, 38, 51
Ivanovski, Dmitri 42

J
journals 10

K
kilopascals 25
Knoll, Max 42
Krebs cycle 16

L
laboratory work 13–14
lichens 59
light microscopes 20, 42
line graphs 27
line of best fit 28
lipid 61
logarithmic scales 29
long answers 15
Lucozade™ 44
lung cancer 6

M
magnification 20–1
mean 22–3
running mean 29
measurements 13
median 22–3
mercury 51
metabolic rate 52
method 10, 11
methylmercury 51
micrograph 20, 40
micrometres 19
microns 12
microscopes 20–1
actual size 21
electron microscopes 21, 42
light microscopes 21, 42
magnification 21
observed size 21
scanning electron microscopes 21
mineral ions 38, 51
mitochondrion 15, 20
mitosis 15, 16
mode 22
models 14
molarity 24
molecules 16
mosses 59
multipotent stem cells 57

N
nanometres 19
Nature 11
Newton, Isaac 4
nitrate 38
nominal data 23
normal distribution 32
nucleic acids 38
null hypothesis 6, 31

O
oestrus cycle 16
open label trials 9
oral glucose tolerance test (OGTT) 44
oral rehydration solution (ORS) 37
ordinal data 23
organisms 16, 46, 50, 57, 59
organs 16
osmosis 25–6

P
parasites 46
Pasteur, Louis 42
pathogens 14, 40
pedigree 48
peer reviews 10–11, 12, 13
percentage 21–22
percentage solutions 25
percentiles 23–24
pesticides 50–51
phosphate 38
phospholipids 38
photosynthesis 38, 50, 59
placebo effect 8
plagiarism 10
plantain 59
platelets 57
pluripotent stem cells 57
pollution 40–1, 50–1
populations 16
potassium 37, 38
pre-implantation genetic diagnosis (PGD) 57
precision 12
prognosis 7
proteins 38

Q
quadrats 29
qualitative data 26
quantitative data 27
quartiles 23–4

R
random errors 12
randomisation 7–8
ratios 22–3
red blood cells 21, 57
references 10
reliability 12
repeatability 12
reproducibility 12, 13
research papers 10
abstracts 10
conclusion 10
data 10
introduction 10
method 10, 11
plagiarism 10
references 10
results tables 26
risk assessment 13–14
running mean 30
Ruska, Ernst 42

S
sampling 30
scanning election microscopes 21
scattergrams 28–9

scientific method 5–6
selective advantage 32
selective breeding 52
serial dilution 25, 40
sewage 40–1
sharing knowledge 10
skills
   accuracy, precision, reliability and validity 12
   animal testing 9
   blind, double-blind and open label trials 8–9
   chi-squared tests 34
   command words 17–19
   concentrations 24
   conclusions 13
   control experiments 7
   correlations and scattergrams 28–9
   drug development 9–10
   ensuring meaning is clear 15
   errors 12–13
   graph types 27–8
   graphs, reading of 28
   Hardy-Weinberg equilibrium 30–1, 48, 49
   hazards and risks 13–14
   investigations involving people 7–8
   line graphs 28
   logarithmic scales 29
   mean, median and mode 22–3
   microscopes and magnification 20–1
   models 14
   normal distribution and continuous variation 32
   organising information 15
   percentages and estimates 21
   percentiles, deciles and quartiles 23
   placebo effect 8
   ratios 22–3
   results tables 26

sampling 29
scientific method 5–6
sharing knowledge 10–12
Spearman rank tests 34, 46
specialist vocabulary 16
standard error 34
standard form 19
statistical tests and values of p 13, 32
statistical tests, choosing the right test 33
statistics 32–5
synoptic essays 15–16
t-tests 35, 52
testable ideas 5
theories, ideas and hypotheses 4–5, 32, 46
units of size 19
units of volume and weight 20
water potential 25–7
writing for your intended audience 15
smoking 6
sodium 37
solute 24
Spearman rank tests 34, 46
specialist vocabulary 16
standard error 34
standard form 19
Stanley, Wendell 42
statins 8
statistical tests 13, 31, 32
stem cells 57
sterile technique 14
synoptic essays 15–16
systematic errors 13
Szent-Gyorgyi, Albert 4

T

t-tests 35, 52
thalidomide 9
theories 4–5
thyroid gland 52
thyroxine 52
tissues 16, 50

tobacco 6
tobacco mosaic 42
totipotent stem cells 57
toxins 42, 50–51, 55
triglycerides 61
trophic levels 50–1
typhoid 37

U

units of size 19
units of volume and weight 20

V

vaccines 8
validity 12
values of p 31
viruses 42
vital ions 37

W

water potential 25–7
white blood cells 57
whole organisms 16, 50, 57
World Health Organization (WHO) 37
writing
   abstracts 10
   audience 14
   clarity 15
   command words 17–19
   conclusion 10
   data 10
   format 15
   intended audience 14–15
   introduction 10
   long answers 15
   method 10, 11
   plagiarism 10
   purpose 15
   references 10
   research papers 10–12
   specialist vocabulary 16

X

X-ray 61

Z

zygotes 57